URBAN ENVIRONMENTAL GOVERNANCE

To my parents,
Alfonso and Eva

Urban Environmental Governance

Comparing air quality management in London and Mexico City

CÉSAR NAVA ESCUDERO
Universidad Iberoamericana (UIA)
Universidad Nacional Autónoma de México (UNAM)

Ashgate

Aldershot • Burlington USA • Singapore • Sydney

Published by
Ashgate Publishing Limited
Gower House
Croft Road
Aldershot
Hampshire GU11 3HR
England

Ashgate Publishing Company
131 Main Street
Burlington, VT 05401-5600 USA

Ashgate website: http://www.ashgate.com

British Library Cataloguing in Publication Data
Escudero, César Nava
Urban environmental governance : comparing air quality management in London and Mexico City. - (The political economy of Latin America)
1. Air quality management - England - London 2. Air quality management - Mexico - Mexico City
I. Title
363.7'392'09421

Library of Congress Control Number: 2001087222

ISBN 0 7546 1564 2

Printed and bound in Great Britain by
Antony Rowe Ltd, Chippenham, Wiltshire

Contents

List of Figures

List of Tables

List of Maps

Preface

Writing a book on urban environmental governance is always accompanied by two unavoidable problems. First, political and institutional changes can develop very rapidly. Indeed, in less than three years (1997-2000), London and Mexico City have experienced tremendous changes in their local government arrangements as new local and democratic units have been set up in both cities. Second, in writing on environment issues one needs to acknowledge that new urban environmental laws and policies develop at an even faster pace. Certainly, diverse urban issues with regard to the problem of air quality management (such as public transport or traffic management regulations), have also been subjected to revision and change in both case studies. This book does not escape the latter, nor does it intend to do so. It was written originally as a doctoral thesis for a PhD at the London School of Economics and was finished in 1998. To my extreme pleasure, those who read it (or at least had a quick glance at it) encouraged me to go further and publish it. Without holding them responsible for whatever the outcome of the book brings, I decided thus to publish it only two years later.

Winds of change in London and Mexico City during the last two years makes me realise that a second edition must come soon. Those of us who have done empirically-based comparative studies know about the intellectual puzzles and financial limitations this represent; this must not however be, an excuse for not working on a new edition. In many ways, I feel indebted to both urban centres: they deserve an updated story of their particular circumstances. For the time being, I hope this book reinforces the argument that 'readers can learn much about urban [environmental] problems by studying them comparatively'.

César Nava Escudero
December 2000

Acknowledgements

Many individuals and organisations have supplied valuable information, given emotional support, and encouraged me in the writing and completion of this book. I would like to begin by thanking Dr. Yvonne Rydin at the London School of Economics for her splendid supervision and enormous patience during the doctoral period. Her continued interest in my progress and her generous support when required, particularly during the last and most difficult stage of the study, is deeply appreciated; I have learned much from her during the whole research process. Special thanks also go to Professor Michael Hebbert at the University of Manchester who supervised this book during its first year and who always showed, from the very beginning, enthusiasm for my work. To him I owe not only my love and curiosity about London, but also the ability to look at this capital city with analytical eyes.

There are too many people who helped me during my fieldwork period in London and Mexico City to cite them all by name, but I would like to mention a few in particular. In London, I would like to thank Mr. Peter Hayden at the London Borough of Havering and Mrs. Yogini Patel at the London Borough of Brent for their time and assistance, so increasing my understanding of many air quality management issues in London. In Mexico City, many thanks go to Mr. Alfonso Aceves Guzmán, Director General at *Dirección de Concertación Política* at the former DDF, for his invaluable support on contacting local authorities across Mexico City. Also, I am grateful to Ms Diana Ponce at the Secretariat for the Environment (SEMARNAP) for opening many doors within different central government authorities. Certainly, I am indebted to all my interviewees whose cooperation (either for the structured survey or semi-structured questionnaires) was crucial to the success of the fieldwork in London and Mexico City.

I also want to express my gratitude to those who kindly gave of their time to read and comment on different parts of the book in draft form, providing positive criticism and assistance with the editing at different stages of the research. These include Prof. Michael Hebbert, Dr. Sylvia Chant, Dr. Cathy McIlwaine, Dr. Andy Pratt, Dr. Judith Cherni, Dr. Alejandro Rodríguez and Ms Susan Dowse. Special thanks go to Dani Bowman de Vera (MA) who kindly accepted the challenge to revise the whole book for proofreading purposes. At the London School of

Economics, I am also grateful to all of those who participated and provided me with helpful comments on different aspects of my book. These took place during research presentations in the Geographical Project Seminar at the Department of Geography and Environment, and in the Mexican-led research workshop organised by Dr. George Philip at the Department of Government.

Many thanks go to Dr. Osvaldo Santin and Juan Osorio who helped me to further my understanding of some of the issues raised here and who also provided me with invaluable friendship, and assisted me in several other ways. I am indebted to Alberto Peredo, Iliana Quiñonez, Iris Hauswirth and Eszter Solti for their words of encouragement during the last year of research. In particular, I wish to thank Dr. Eduardo Torres at UNAM for the support given to make this book a reality, and to Rubén Muñoz at UIA for giving permission to revise and edit the final manuscript of the book. For technical and computing assistance I am grateful to Daniel Gutiérrez Carreón, and in particular, to Martha Estrada at SEMARNAP who never said 'no' to my vast amount of computing queries always showing enormous patience.

I would also like to thank the Mexican government for its financial aid; this research was made possible by a *Consejo Nacional de Ciencia y Tecnología* (CONACYT) postgraduate loan.

Finally, there are very few occasions in which one can express gratitude to family members. I take this opportunity to thank my parents and my two brothers who, no matter what, have been always there to give their unconditional support and invaluable understanding.

List of Abbreviations

ALA	Association of London Authorities
ALDF	*Asamblea Legislativa del Distrito Federal*
ALG	Association of London Government
AMA	Association of Metropolitan Authorities
AQMS	Air Quality Management System
ARDF	*Asamblea de Representantes del Distrito Federal*
AUN	Automatic Urban Monitoring Network
BATNEEC	Best Available Techniques Not Entailing Excessive Cost
BPM	Best Practicable Means
BR	British Railways
CAM	*Comisión Ambiental Metropolitana*
CEC	Commission for Environmental Cooperation
CED	Committee for Economic Development
CEE	Commission of the European Communities
CFE	*Comisión Federal de Electricidad*
CIEHO	Chartered Institute of Environmental Health Officers
CLEAR	Campaign for Lead-Free Air
CMPCCAVM	*Comisión Metropolitana para la Prevención y Control de la Contaminación Ambiental en el Valle de México*
CO	Carbon monoxide
CO_2	Carbon dioxide
Constitution	*Constitución Política de los Estados Unidos Mexicanos*
CROM	*Confederación Regional de Obreros Mexicanos*
DDF	*Departamento del Distrito Federal*
DETR	Department of the Environment, Transport and the Regions
DF	*Distrito Federal*
DoE	Department of the Environment
DoH	Department of Health
DoT	Department of Transport
DTI	Department of Trade and Industry
EA	Environment Acti 1995
EC	European Community
EdoMex	*Estado de México*
EEC	European Economic Community
EHO	Environmental Health Officer
EPA	Environmental Protection Act 1990

EU	European Union
EUN	Enhanced Urban Network
FoE	Friends of the Earth
GDF	*Gobierno del Distrito Federal*
GLA	Greater London Authority
GLC	Greater London Council
GoL	Government Office for London
GTZ	German monitoring agency
HC	Hydrocarbons
HMIP	Her Majesty's Inspectorate of Pollution
HMSO	Her Majesty's Stationery Office
ILEA	Inner London Education Authority
IMECA	*Indice Metropolitano de la Calidad del Aire*
IMP	*Instituto Mexicano del Petróleo*
INE	*Instituto Nacional de Ecología*
INEGI	*Instituto Nacional de Estadística, Geografía e Informática*
INSP	*Instituto Nacional de Salud Pública*
IPC	Integrated Pollution Control
JICA	Japanese International Cooperation Agency
JLAP	Joint London Advisory Panel
LADF	*Ley Ambiental del Distrito Federal 1996*
LAPC	Local Air Pollution Control
LAQN	London Air Quality Network
LBA	London Boroughs Association
LCC	London County Council
LDDC	London Docklands Development Corporation
LFCDA	London Fire and Civil Defence Authority
LFPA	*Ley Federal de Protección al Ambiente 1982*
LGA	Local Government Act 1985
LGEEPA	*Ley General del Equilibrio Ecológico y la Protección al Ambiente 1988*
LOAFDF	*Ley Orgánica de la Administración Pública del Distrito Federal 1994*
LODDF	*Ley Orgánica del Departamento del Distrito Federal 1978*
LPAC	London Planning Advisory Committee
LRB	London Residuary Body
LRC	London Research Centre

LSS	London Scientific Services
LT	London Transport
$\mu g\ m^3$	Micrograms per cubic metre
M25	London's orbital motorway
MAMC	Metropolitan Area of Mexico City
MBW	Metropolitan Board of Works
MZMC	Metropolitan Zone of Mexico City
NAFTA	North American Free Trade Agreement
NAQS	National Air Quality Strategy
NETCEN	National Environmental Technology Centre
NGOs	Non-governmental organisations
NO_2	Nitrogen dioxide
NO_X	Nitrogen oxides
NSCA	National Society for Clean Air
OECD	Organisation for Economic Cooperation and Development
OPEC	Organisation for Petroleum Exporting Countries
OPSIS	Real time monitoring system
O_3	Ozone
PAHO	Pan American Health Organisation
PAN	*Partido Acción Nacional*
Pb	Lead
PEMEX	*Petróleos Mexicanos*
PGs	Process guidance notes
PICCA	*Programa Integral Contra la Contaminación Atmosférica de la Zona Metropolitana de la Ciudad de México*
PM_{10}	Particles less than 10μm
Ppm	Parts per million
PRD	*Partido de la Revolución Democrática*
PRI	*Partido Revolucionario Institucional*
PROFEPA	*Procuraduría Federal de Protección al Ambiente*
PT	*Partido del Trabajo*
PVEM	*Partido Verde Ecologista de México*
QUANGOS	Quasi-governmental organisastions
QUARG	Quality of Urban Air Review Group
RAMA	*Red Automática de Monitoreo Atmosférico*
RCEP	Royal Commission on Environmental Pollution
RCLG (1960)	Royal Commission on Local Government in Greater London 1957-1960

RCLG (1969)	Royal Commission on Local Government in England
RIAPDF	*Reglamento Interior de la Administración Pública del Distrito Federal 1995*
SCT	*Secretaría de Comunicaciones y Transportes*
SE	*Secretaría de Energía*
SECOFI	*Secretaría de Comercio y Fomento Industrial*
SEDESOL	*Secretaría de Desarrollo Social*
SEDUE	*Secretaría de Desarrollo Urbano y Ecología*
SEIPH	South East Institute of Public Health
SEMARNAP	*Secretaría de Medio Ambiente, Recursos Naturales y Pesca*
SEP	*Secretaría de Educación Pública*
SERPLAN	South East Regional Planning Conference
SHCP	*Secretaría de Hacienda y Crédito Público*
SO_2	Sulphur dioxide
SPM	Suspended particulate matter
Sq. Kms.	Square kilometres
SSA	*Secretaría de Salubridad y Asistencia*
Ssa	*Secretaría de Salud*
STC	*Sistema de Transporte Colectivo - Metro*
STI	*Secretariado Técnico Gubernamental*
Tons / yr	Tons per year
TPPs	Transport plans and programmes
TSJDF	*Tribunal Superior de Justicia del Distrito Federal*
TSP	Total suspended matter
TV	Television
UCL	University College London
UK	United Kingdom
UN	United Nations
UNAM	*Universidad Nacional Autónoma de México*
UNCED	United Nations Conference on Environment and Development
UNCHE	United Nations Conference on the Human Environment
UNCHS	United Nations Centre / Conference for Human Settlements (HABITAT)
UNEP	United Nations Environment Programme
US / USA	Unites States of America
USEPA	Unites States Environmental Protection Agency
VOCs	Volatile Organic Compounds
WB	World Bank

WCED	World Commission on Environment and Development
WHO	World Health Organisation
WRI	World Resources Institute

1 Introduction

> *One of today's 'buzz' words is 'megacity' (over 10 million), and Mexico City with almost 15 million [inhabitants] in 1990 is certainly mega-big...if one is sensible and reasonably sensitive, then it is not a particularly dangerous city. Nor is it an easy city in which to live permanently. People often challenge me and ask, 'Would you want to live permanently in Mexico City?. 'Hell, no!', I reply, but then nor do I want to live in London, New York or any other large metropolitan area where many of the same hassles and problems apply.*
>
> *Peter Ward (1990) in the first edition of his book 'Mexico City'*

Rationale of the book

During the last decade London and Mexico City have suffered from perceived poor air quality due to increased concentrations of road transport emissions. Current data on mortality rates show that in 1991 a wintertime pollution episode in London contributed to 100-180 premature deaths; similarly, it is said that high levels of pollution in Mexico City during 1990 contributed to an estimated total number of 6 400 excess deaths. In addition, the number of health studies on morbidity in these two cities shows that air pollution has produced adverse effects on the health and well-being of their inhabitants. Managing air quality[1] has become therefore an issue of concern for both urban centres. In the process of governing London and Mexico City local authorities play a crucial role in addressing and tackling this urban environmental problem, respectively.

This book is an exploration of urban environmental governance[2] issues. A major reason being that until now, research within environmental studies has concentrated on the preservation of natural resources or the 'Green Agenda' and has recently turned to recognise urban environmental issues or the 'Brown Agenda' (Gilbert et al 1996: 13-14; Serageldin and Cohen 1995: 1). The importance of urban areas at the fore of the environmental debate has been increasingly recognised due to the positive and negative effects they have on the natural environment (see Breheny 1992: 277; Haughton and Hunter 1994: 1; Stren 1992: 1). While the need to adopt a comprehensive or integrated approach to urban environmental problems means including key aspects such as public policies, political will, private sector involvement, social participation, government intervention at all levels, and so forth, this research addresses just one of these variables: local government intervention. The reason for this is because local governments, increasingly embedded now as one of the key components of local governance, have been recognised as crucial for promoting, designing and implementing urban environmental policies as well as acting as mechanisms for enhancing individual participation (Blowers 1993b: 16-17; Gilbert et al 1996: 24-28; Gordon 1993: 13; Johnston and Pattie 1996: 672; UNCHS 1996a: 161). Having said this, the identification of the way in which local governments intervene in urban centres becomes crucial. That is to say, the manner in which local governments are organised needs to be addressed and revised (see Haughton and Hunter 1994: 300-303; OECD 1990: 38-43; WCED 1987: 243-247).

Reviewing the organisation of local government has traditionally implied analysing such aspects as the organisational structures or units of government, functional allocation, democracy and accountability, financing of local authorities, and the question of central-local government relations (see, for example, Barlow 1991: 10; Rhodes 1980: 574). The organisational structure issue (particularly in terms of size and boundaries) has been given special consideration among local government organisation studies. So, for example, according to Humes and Martin (1969: 34) the structure of local government is relevant in as much as it represents 'the framework within which local [and central] public policy is determined and implemented'. Furthermore, it has been argued that:

> the structure is perhaps the most important determinant of the nature of local government and administration and of the way in which it functions (Leemans 1970: 51).

Over the last years, it has been similarly stated that devoting more attention to diverse aspects of local government, especially regarding organisational structures, would significantly contribute to the understanding and designing of solutions to the problems of local governance (see, for example, Humes IV 1991: xi; Rowat 1980: xv). Discussing the organisation of local government therefore needs to consider both the units or structures themselves and the other subjects of traditional studies of local government, such as finance, functions, governmental interrelations, accountability, and so on (Humes and Martin 1969: 34-35; Leemans 1970: 29; Rhodes 1980: 572-574).

This research focuses on two main areas of current local government organisation studies. First, it concentrates on the description and analysis of what academic literature has identified as organisational structures (also known as institutional arrangements or units of local government) and other traditional aspects (particularly functional allocation, and intra- and inter-governmental relations) (see, for example, Barlow 1991; Bish and Ostrom 1973; Gunlicks 1981; Hesse and Sharpe 1990/91: 603-604; Humes and Martin 1969; Leemans 1970; Wolman 1995). Second, it also concentrates on the analysis and description of key actors that in the past were not considered or simply were not actually involved in the process of governing metropolitan areas. These other actors include special-purpose or ad hoc authorities, non-governmental organisations, the private sector, and diverse forms of public participation. The reasons for including these actors in today's local government studies, are due to the changing conditions and values in society and because of the current importance of the emerging issues on urban environmental governance (Gilbert et al 1996: 16; Humes IV 1991: xi; Jones and O'Donnell 1980: 541; Serageldin and Cohen 1995: 12-15).

While discussion on the organisation of local government has been the subject of much debate in developed countries, it has only recently become evident in developing countries (Leemans 1970: 13; McCarney 1996d: 5; Rowat 1980: 375). In the North, the debate on how to organise local government has developed, particularly in the United States (USA) and in the United Kingdom (UK), in this last century. The traditional discourse on reforming local government - mainly concentrated on organisational structures - has focused on the issues of size and area of local government (see Barlow 1991: 8-10). In the United States, for example, the debate on improving urban government in the last century was dominated by the centralised views and recommendations put forward by the consolidationist school - also known as the reform movement - which advocated large scale units of government (Gunlicks 1981: 11). It was not until the late 1960s, and particularly during the 1970s, that a decentralised and alternative logical-deductive model - the public choice theory - started to use concepts and methods from economics in order to explain urban governments' behaviour. This theory has focused on individual preferences and on the diverse nature of goods and services rather than on the organisation structure (Bish and Ostrom 1973: 17; Dunleavy 1991: 147). In Britain, the academic debate on favouring large scale units of government - initiated by William Robson (1939) -

blossomed during the post-war period. The emerging 'New Right' ideology of the 1970s, though, started challenging the idea of large-scale or metropolitan government. This ideology favoured political fragmentation whereby local communities and the private sector would widen their participation (see John 1990: 18-19; King 1995: 236-238).

During the 1980s and 1990s, the debate on reforming the structures of local government has continued. There are two trends, though, that seem to characterise these recent years of debate among Western democracies. First, although there seems to have been a retreat of the idea of metropolitan government, it has been argued that the metro model is far from dead and that it has actually come back onto the political agenda, especially in countries like the UK (see Barlow 1991: 36; Keating 1995: 118; Sharpe 1995b: 22). Second, the debate on confronting diverse models of government during the last decades - i.e. consolidation vs. fragmentation - has increasingly started to shift towards new avenues of discussion which are closely related to the issues of local governance and urban sustainability. The latter has been manifested through a change in the focus of debate over the role local authorities currently have. So, while within local government and governance studies there is an increased call for local authorities to think of themselves generally as 'enablers' rather than as direct providers of services, within urban sustainability studies the recommendations are similarly for local authorities to increasingly act as enablers, coordinators and facilitators of environmental strategies and policies (Haughton and Hunter 1994: 300; Wolman 1995: 153-158).

In countries of the South, the debate on reforming local government also developed during the post-war period, but at a very slow pace. It is actually much more difficult to generalise in less developed countries on the type of debate surrounding local government organisation. As explained by Rowat (1980: 376), because developing countries are spread around the world and have different histories with diverse conditions, a generalisation of the nature of their reorganisations becomes difficult to make. In spite of this, it is possible to depict some common characteristics from a historical point of view. For instance, due to Western imperialism and colonialism diverse systems of local government of the North have been imposed in developing countries over the last centuries. As countries in the South became independent, they have either inherited the same colonial system of local government or tried to introduce new structures as a result of changed political conditions (see also Leemans 1970: 13). In many cases, the reform of local government has taken place because a new state system of government at central level has been adopted. The latter, in turn, has often been a copy - frequently distorted - of other existing models of central, and consequently, of local government in the North. Overall, local government has been a neglected tier of government in the development effort (McCarney 1996d: 5).

The organisation of local government in the South has often been reviewed either by region - Africa, North America, South America, Asia (see, for example, Alderfer 1964; Humes and Martin 1969; Mawhood 1993; McCarney 1996a and 1996c) - or by system of local government - French, Germanic, Soviet, British models (see, for example, Humes IV 1991). Very few have actually reviewed local government organisation by distinguishing them within the unitary and federal systems of state government dichotomy (see Rowat 1973).

The origins for the discourse on local government organisation in countries of the South has not been the issue of scale-enlargement as it has in the North. Rather, the debate on local government reform has focused on the political reliability of the system itself in terms of social and economic development, democracy and public participation, and political rights (see Akin Aina et al 1991: 4-6; Leemans 1970: 13). Although some debate on organisational structures has actually taken place at metropolitan level, there exists an enormous gap in the academic literature of this field. In Latin America, for example, the debate on the general issue of local government organisation itself is scarce - let alone on the detail of structural arrangements. This is partly because the issues of democracy and political rights in urban centres have overshadowed those of city management, and partly because large or mega cities started to appear only during the 1960s and 1970s (see Richardson 1993: 48-54).

Within the debate on local government organisation in countries of the North and of the South, seldom has there been an attempt to revise diverse proposals or models for reform in the context of the emerging urban environmental demands of the 1990s. This gap in academic literature exists despite the fact that much of the understanding and scientific progress on local government can develop from doing comparative studies; yet, such studies have not been recurrently done (see Fried and Rabinovitz 1980: 19; Hesse and Sharpe 1990/91: 603; Humes IV 1991: x-xi; Rhodes 1980: 563; Rowat 1980: xiii). Hence, this book seeks to foster the understanding of urban government in developed and less developed countries by comparing two different local government organisations: Greater London and Mexico City. In doing so, it also aims at reducing the existing gap on comparative local government organisation studies. In order to achieve the latter, the book carries out a pragmatic, empirically-based-analysis of the two localities in the context of their own governmental systems. Given the differences between each city, key governmental characteristics (e.g. constitutional status, democracy, accountability, ad hoc agencies, and so on) pose some difficulties for comparative analysis as they have diverse connotations and play different roles in each of them. Nevertheless, while comparing local government has always represented analytical puzzles, especially within a North-South context, the scale of these two world cities brings some generic problems of local administration and urban environmental issues. As Fried and Rabinovitz (1980: 19) argue 'readers can learn much about urban problems by studying them comparatively.' The theoretical justification for comparing London and Mexico City is because the former is currently a system of government with a simple, single tier of 33 unitary authorities with no city-wide authority, and the latter consists of a highly centralised, city-wide and mayor-headed local authority. Analysing these two different systems of urban government is, in itself, a major justification for doing comparative studies and thus advancing the understanding of local government systems.

Among the existing urban environmental problems in cities, the book focuses on urban air pollution. This is because air pollution has been a significant environmental problem in urban centres over the last decades. According to Elsom (1996: 1) urban air pollution is currently one of the major problems in developed and developing countries as it threatens the health and well-being of about one-half of the world's urban population. In fact, the United Nations Centre for Human Settlements HABITAT (UNCHS 1996a: 143) has categorised air pollution, together with water pollution, the collection and management of solid wastes, and noise pollution, as one of the four most serious city-wide environmental problems.

Much of the existing literature on air quality management issues in urban centres has actually concentrated on describing the cases of diverse cities in different countries but has not always included a serious empirically-based comparative approach. Among the few comparative exercises at a worldwide or regional level some of the most important are Bennet (1991); Medina and Quenel (1993); Murley (1991); OECD (1995); WHO/UNEP (1992). Other relevant studies have included two, three or four specific case-studies, such as Bridges (1995); Elsom (1996);

Eskeland et al (1994); Levinson and Shetty (1992); UN (1984). The subjects of study within such comparative frameworks have mainly addressed policy measures and air pollutants' effects on human health as well as air quality standards, monitoring systems, legal regimes, and statistical information. None the less, little has been done concerning how a particular type of local government organisation within the process of urban governance can contribute to improve such a problem.

The analytical framework

The purpose of this book is to combine an analysis of the impact of political and institutional changes for air quality management looking at two case-studies: London and Mexico City. By adopting a comparative approach, the analysis focuses on how diverse local government structures operate in relation to air pollution control from the perspective of local and central authorities and other key agencies. In so doing, the book outlines the debates on reorganising local government by focusing on three perspectives which offer diverse explanations to urban environmental governance problems: the orthodox public administration model, the public choice theory, and the local governance approach. This is because this book attempts to distinguish the advantages and disadvantages of each model for reorganising local government in order to improve air quality. The main argument considers the need for both an upper-tier area-wide coordinating authority and lower-tier politically fragmented government units at local level for improving air quality in both urban centres.

The discussion and analysis of the empirical data presented in this book[3] (Chapters 6 and 7) are based on qualitative methods following a technique known within the social sciences as intensive research design (see, for details, Harré 1979: 132-134; Sayer 1992: 241-251; Sayer and Morgan 1985: 150-157). In order to facilitate the comparative analysis, the qualitative method followed in London and Mexico City was as similar as possible for both cases. This method consisted in elaborating and carrying out semi-structured questionnaires (face-to-face interviews) with central and local government authorities, non-governmental organisations and other key policy makers involved in air quality management. In the case of London, the arguments in chapter six are developed from the analysis of thirteen semi-structured interviews with key policy makers carried out in London during the period November 1994 - May 1995. The latter is complemented by four updating or second interviews carried out during the years of 1995 and 1996 and by a number of informal discussions with other air quality management participants. Additionally, several visits to air quality manual and real time monitoring sites were included as part of the fieldwork. Likewise, in the case of Mexico City, the arguments in chapter seven are developed from the analysis of sixteen semi-structured interviews with key policy makers carried out in Mexico City during the period May - September 1995. As with the case-study in London, the latter is complemented by two updating or second interviews carried out during the years of 1996 and 1997 and by a number of informal discussions with other air pollution control participants. Several visits to air quality real time monitoring sites were also included as part of the fieldwork.

Table 1.1 Local authorities interviewed through structured questionnaires

	London	*Mexico City*
Interviews	29 boroughs + City of London	16 *delegaciones*
No response	4 boroughs	-

Source: Author's survey

In order to complement the analysis of the semi-structured questionnaires, the research also uses a technique known within the social sciences as extensive research design (see Harré 1979: 132-133; Massey and Meegan 1985: 145; Sayer 1992: 241-251). The main aim of this was to increase and collect further data on air quality management issues through structured questionnaires (i.e. surveys) which targeted local authorities at the lower-tier level - boroughs in London and *delegaciones* in Mexico City. In both cities, the surveys were sent to local authorities involved in environmental or pollution control issues. In the case of London, responses were obtained from twenty nine borough officers such as Environmental Health Officers, Pollution Inspectors, Senior Technicians (Pollution Sections), Built Environment Director, and so on. In Mexico City, the completed questionnaires were obtained from all sixteen *delegaciones* that include Heads of Environmental Offices or Administrative Units, Urban Development Units, Social and Cultural Services Offices, and so forth (see Table 1.1). The analysis of the structured questionnaires is presented in Chapter 5.

I acknowledge that some methodological limitations exist, not only those that are commonly found in other investigations that use intensive and extensive techniques, but particularly with regard to the many complex analytical problems of assessing the meaning of questions and responses. It is not the aim of this book however, to discuss how to conduct and analyse an interviewing process - i.e. through a stimulus response model, discourse method, and so on. The original purpose of the interviewing is:

> to understand what respondents mean by what they say in response to our queries and thereby to arrive at a description of respondent's worlds of meaning that is adequate to the tasks of systematic analysis and theoretical interpretation (Mishler 1986: 7).

Thus, the views of the interviewees expressed and analysed in the book are treated as causally related individual stories. The understanding, analysis, and explanation of these stories is ascertained in each case according to the particular individual's context in terms of their institutional and political interests regarding air pollution control in London and Mexico City, respectively. Furthermore, the interviewees are seen as policy actors whose responses are closely linked to their own personal benefits if and when policy or institutional changes take place. The creation of the structures or categories for these explanations is based on my own understanding of the issues according to the theoretical framework presented in the book. It must be noted that the stories presented in chapters five, six and seven are not necessarily identified with the official or authoritative views of the agencies or governmental units that the interviewees represent.

An overview

This book is divided into eight chapters. Chapter 1 gives a general view of the book introducing the reader to its rationale and analytical framework. The second chapter provides a descriptive and analytical insight into the main environmental issues that are addressed in the book. The aim of this chapter is twofold. First, it gives an introductory summary of the most relevant events that have taken place within the environmental debate (focusing on the urban environmental agenda) in countries of the North and the South of the last decades. Second, it centres on the topic of the book: urban air pollution. In so doing, it briefly examines the origins and consequences of air pollution at the local, national and global levels, as well as portraying the scale of the problem in the two case-studies: London and Mexico City. The third chapter examines the role of local governments within an urban environmental context as well as considering the theoretical debates on reforming their organisational structures. The argument centres on the need for an area-wide coordinating authority at the local government level (without excluding lower tier authorities' participation) for dealing with air quality management issues. The main purpose of this chapter is twofold. First, it stresses the importance of local governments as essential components for governing urban centres. Second, it outlines the debates on reforming local government by focusing on three different perspectives: the orthodox public administration model, the public choice theory, and the local governance approach. The main objective in this section is to examine the advantages and disadvantages of each model for reorganising local government to bring about improved air quality.

Chapter 4 has two main purposes. First, it aims to bring into a comparative framework the main local structural arrangements in London and Mexico City. In doing so, it describes and analyses the current organisation of local government in each case-study by examining a number of variables such as geographical areas, population, functional allocation, finance, democracy and accountability, and so on. Second, it discusses variants of local government systems by distinguishing the two case-studies. This part seeks to identify from a historical perspective both the debates and experiences on governmental reform that London and Mexico City have undergone over the last decades, and the relationship of their government structures with air pollution control. While highlighting the metropolitan government response to air pollution, the main purpose of this chapter is to pull the two cities into a single scheme of interpretation. The fifth chapter brings into comparative perspective the views of London and Mexico City's local authorities regarding local government organisational structures and air quality management issues. This chapter analyses in a comparative fashion various air pollution and local government issues based on a survey (structured interviews) carried out with local authorities in both urban centres. It also seeks to review the lessons that can be learned from each case-study according to their systems of local government and air quality management strategies in the light of the three approaches outlined in chapter three. The sixth and seventh chapters examine the views of central and local government authorities, and other key actors (e.g. non-governmental organisations) on the organisational structures of local government in London and Mexico City regarding air pollution control, respectively. The analysis of how air quality management works in both cities include such issues as air quality monitoring systems, transport and traffic management, and the application of an eventual emergency plan. This chapter seeks to make reference to the relevance of the empirical information provided to the three approaches of local government. The final chapter gives an overall view of the main findings in this book. A brief section on further avenues for research is included in this last chapter.

Notes

1 The terms 'air quality management' and 'air pollution control' are used interchangeably in the vast academic literature on air quality. Likewise, this book uses these two terms interchangeably. Nevertheless, it is important to note that 'air quality management' is the currently preferred policy term.

2 For an explanation of the term 'governance' at city level, see Chapter 3.

3 For a more detailed explanation of the methodology used in this book (i.e. the intensive and extensive methods, the questionnaires, methodological limitations, and so on), see Nava-Escudero (1998).

2 The Environmental Debate and the Problem of Urban Air Pollution

...they [my ancestors] told me that there used to be trees, butterflies, that the air was pure, the rivers crystalline; they told me that in the past people were humble and knew the language of flowers, and also sang at sunrise and danced at sunset...[but]...when I opened my eyes everything had gone: and the only thing I could feel, I could be certain of was that our ancestors spoke a common language, which the trees speak during the day, and the stars during the night.

Chamalú
(Quechua Indian and Andean Shaman)

Introduction

Over the last few decades, environmental issues around the world have gained increasing importance within political agendas at the international, national, and local levels (Mannion and Bowlby 1992: 17; Thomas 1992: 61). The perception of a current environmental crisis stems from the concerns during the 1960s and 1970s - particularly in countries of the North - over the depletion of natural resources, the pollution of the environment and the associated health impact (Lovell and Johnson 1994: 199; Rees 1991: 1; White 1993: xi-xii; Wilson 1984: 3). The predominant worldview during those two decades concerned local or regional environmental problems and resource depletion. This shifted during the 1980s to a debate on global and common sustainability concerns. During the first half of the 1990s, the environmental debate started to focus on urban environmental concerns, particularly on the issue of 'sustainable cities' or 'sustainable urban development'. Many older concerns with public health in cities have now been subsumed within this urban sustainability agenda. In this discussion much attention has been paid since then to the need for diverse forms of local action in order to improve environmental conditions in urban centres. This focus on local practice has in turn recognised the fundamental role that local governments can play in tackling city-wide pollution. Strengthening the participation of local authorities as well as creating more efficient, effective, coordinated, and representative local government structures have been highlighted as key issues that need to be addressed when dealing with urban environmental issues (see Gilbert et al 1996: 23-42; Hardoy et al 1992: 162-164; Haughton and Hunter 1994: 300-303).

As well as giving a summary of some of the principal events and views that have emerged from the environmental debate in countries of the North and the South, the aim of this chapter is twofold. First, it highlights the need for local action, particularly by local authorities, in dealing with urban environmental issues. Second, it focuses on the main environmental issue that is addressed in this book: urban air pollution. In so doing, it briefly examines the origins and consequences of this particular issue at the local, national and global levels as well as portraying the scale of the problem in the two case-studies: London and Mexico City. The choice of two cities across the North-South divide necessitates some consideration of the different forms the environmental debate has taken in the North and the South. This is the focus of the next section.

Environmental debates: the north-south distinction

The terms 'North' and 'South' usually represent developed and developing countries. Within this terminology, it is generally accepted that the 'North' consists of North America (USA and Canada), Europe (Western), the former Soviet Union, Japan, Australia, and New Zealand; the 'South' includes the rest of the world's countries (Pearson and Pryor 1978: 3-6; White 1993: 3). This distinction between developed and developing countries, generally based on economic income grounds (see Keating 1993: ix), can help us to understand the diverse origins and views that have evolved within a rapidly expanding environmental discourse.

The environmental debates of the late 1960s and early 1970s in countries of the North, originated from concerns about preserving the natural landscapes, the impact of toxic chemicals, and in particular, from resource depletion (White 1993: 12). The issue of scarcity of resources - which became a major issue of concern with the first OPEC oil shock (Sandbrook 1984: 12) - led to the discussion on 'The Limits to Growth' thesis of the Club of Rome which linked projections of population, resources and economic growth (see Meadows et al 1972; White 1993: 14). This thesis centred on the Malthusian argument that uncontrolled population growth and resource exploitation would be the cause of an imminent exhaustion of humans' stock of natural resources. The environmental debate focused on the incompatibility between continued economic growth and environmental protection (see Ehrlich 1969; Ehrlich and Ehrlich 1970; Gandy 1993: 10; McCormick 1995: 80-81; Meadows et al 1972). However, it was not until the North realised that environmental issues were 'indeed global and that concern for environmental quality was not exclusively the domain of the rich' (Pearson and Pryor 1978: 1), that it began to recognise the need for including the South within the emerging environmental movement. It then rapidly became apparent that there was a distinction in the concerns of the North and the South. After the 1972 UN Conference on the Human Environment (UNCHE) and the Cocoyoc Declaration of 1974,[1] it was acknowledged that environmental concerns in the South were not primarily about the conservation of the countryside, the use of chemical toxics, or resource depletion. Rather, the immediate environmental interests of the South were related to development problems; that is to say, there was an emphasis on the relations between the environment and development. The position of the South toward environmental issues has therefore included the 'basic needs' and 'strategies to survive' arguments, as well as socio-cultural and environmental issues such as equity, cultural diversity, self-reliance of communities, democracy, and participatory and self-management resources (Atkinson 1991: 406-410; Leff 1991: 134-135 and 1995: 120; Pearson and Pryor 1978: 2-3; Redclift 1984: 46-47; Redclift and Goodman 1991: 4). This led various countries in the South to consider the need for a strategy in which an environmental dimension would be inserted into the development process. This strategy, which emerged during the 1970s from UNCHE and the newly created United Nations Environment Programme (UNEP), became widely known as 'ecodevelopment' (see Brañes 1994; García Guadilla 1991; Székely 1978). In countries of the South, while the idea of ecodevelopment continued to influence the environmental debate of the 1980s and 1990s, it was subsumed within the new incoming slogan 'sustainable development' in the early 1980s (Leff 1995: 59; MacManus 1996: 50). Meanwhile, the pessimistic neo-malthusian thesis of the North rapidly became the subject of much criticism and the environmental debate during the 1980s and this shifted attention from the concerns of scarcity of resources to the environmental consequences of using those resources. By the 1990s, the discussion in countries of the North had also moved on to the issues of sustainability and development (Gandy 1993:

13; O'Riordan 1981: 60-65; Soussan 1992: 22-23).

Following the 1987 Brundtland Commission's Report 'Our Common Future' (see WCED 1987), in 1989 the UN called for a worldwide conference - the United Nations Conference on Environment and Development (UNCED) - which took place in Rio de Janeiro in June 1992. Considered as a 'milestone event', UNCED (also known as the 'Earth Summit') raised consciousness on global issues such as the loss of tropical forests and of the world's biological diversity, the changing climatic conditions, and particularly on the issue of sustainable development (see Birnie and Boyle 1992: 4-5; Grubb et al 1993: 6-7; Keating 1993: 6). Although the interests and attitudes of the North and the South became more conciliatory in the 1980s (McCormick 1995: 179), their environmental concerns and priorities even under the umbrella of sustainable development remained dissimilar.[2] By the end of the 1980s and early 1990s, the environmental concerns of the North were about carbon dioxide emissions, ozone depletion, loss of rare species and toxic waste disposal. For the South it was about basic needs: poor water supply, bad sanitation, soil erosion, depletion of wood supplies and environmental health. At the same time, the South highlighted the importance of reducing the burden of debt to Northern banks and governments in order to better address environmental problems. In addition, it became clearer that the North had a great deal of shared responsibility - perhaps the dominant responsibility - for issues such as global warming (see Redclift and Sage 1994: 4-8).

During the 1990s, the environmental debate in countries of the North and the South started to focus on urban environmental issues and shifted toward the notion of 'sustainable urban development'. The increasing debate over sustainable urban development issues within the North-South framework provided for a distinction between the 'Green' and the 'Brown' Agendas. The 'Green Agenda' - identified as a Northern challenge - usually refers to the preservation and management of natural resources focusing on issues such as resource depletion, deforestation, biodiversity and global warming. Conversely, the 'Brown Agenda' - identified as a Southern challenge - commonly refers to urban environmental issues in developing countries concentrating on the health impact that derives from inadequate water, sanitation, drainage and solid waste services as well as poor waste and air quality management (see Bartone et al 1994: 1; Serageldin and Cohen 1995: 1; WB 1994: 32). Despite this distinction, it is clear that 'brown' issues are not purely within the domain of developing countries. Although cities in the North often have the means to reduce the pollution they produce locally, they can also suffer from severe urban environmental problems such as air pollution or waste disposal (Gilbert et al 1996: 13). Likewise, not all urban centres in the South, like the rapidly growing city of Curitiba in Brazil, suffer from serious environmental problems (Hardoy et al 1992: 17). In the case of air pollution (which is the main focus of the book) achieving healthy urban air quality is not only a local indicator of sustainability in urban centres but a matter of much local concern for countries in the North and the South. Indeed, among the existing environmental problems in urban centres, air pollution has now been categorised as one of the most serious urban-wide problems in developing and developed countries (Elsom 1996: 1 and 214-220; Hatcher 1996: 192-197; UNCHS 1996a: 143).

The idea of achieving urban sustainability began to address not only the environmental but also the socio-economic, cultural, political and institutional aspects of the city (see Breheny 1992a: 2 and 1992b: 277; Greig 1993: 1; Hardoy et al 1992: 171; Haughton and Hunter 1994: 10; Stren 1992a: 1). In other words, securing sustainability at the urban level was seen as involving the adoption of a comprehensive or integrated approach to urban environmental problems. An integrated strategy requires the inclusion of key aspects such as public policies, political will, private sector involvement, social participation, government intervention at all levels, and so forth (see Breheny and Rookwood 1993: 150; Elkin et al 1991: 9; Hardoy et al 1992: 188-197; Keating 1993: ix; OECD 1990: 39-43 and 1996: 168; WCED 1987: 243-247). The reasons for bringing about sustainable urban development by adopting an integrated approach, though, are not necessarily the same for developed and developing countries. By way of illustration, for those cities with high levels of non-renewable resource use - usually wealthy cities - the priority is to reduce levels of fossil fuel use and waste generation, while maintaining a productive economy and preserving the natural and built environment. For cities with low levels of non-renewable resource use - usually poor cities - the priority is to achieve socio-economic and political goals preserving the natural and built environment without increasing that resource use. Furthermore, although cities in the North and the South face the challenge of reducing local pollution, the North also sees urban sustainability as a major contributor to the solution of global problems, and the South mainly sees it as an urgent need for solving local problems (see Gilbert et al 1996: 14; Hardoy et al 1992: 188-189).

It is within the context of UNCED (in Chapter 7 of Agenda 21) and the debate over sustainable urban development that the need for local action and the fundamental role of local authorities started to be addressed. The latter also gained greater official recognition with the second UN Conference on Human Settlements (UNCHS) - Habitat II, also known as the 'City Summit' - held in Istanbul, Turkey, in 1996 (see Satterthwaite 1997: 1686; Serageldin and Cohen 1995: 1). This conference resulted in a broad consensus on the fact that the future of the Earth will depend to a great extent on the quality of life in urban centres. It was ascertained that both the scale and scope of the issues, and the participants concerned with human settlements have changed and actually expanded since Habitat I - the first UN Conference on Human Settlements held in Vancouver, Canada, in 1976. Habitat II addressed the issues and problems of human settlements (urban and rural) into the next century including such new topics as democracy, human rights, participation, sustainability, government decentralisation, women's empowerment, and public-private partnership. Additionally, it included the participation of mayors and representatives of local governments who had the opportunity to speak at conference sessions, and NGOs which actively intervened through organised events, demonstrations and diverse activities (see Carlson 1996: 4-5; Cohen 1996: 8; Hundsalz 1996: 6; UNCHS 1996b: 1).

The emerging focus on the participation of local authorities in urban centres has led to a concern that most governments at local level (particularly in the South) lack both the trained personnel and the financial base to control pollution, as well as the institutional means to ensure an effective, coordinated and representative governmental response to environmental problems. Although the South's urban environmental problems may be wider in scope and more traumatic and human health-damaging than in the North, allocation or devolution of responsibilities for environmental management to local level has occurred in both developed and developing countries (see Gilbert et al 1996: 24; Hardoy et al 1992: 220). By way of illustration, in both the case studies presented in this book - London and Mexico City - there is evidence of shifting responsibilities on air quality management matters from central to local government, for example, regarding air quality monitoring systems (see Chapters 5-7).

While some differences exist within a North-South context for addressing urban environmental issues, recommendations for reorganising the structures of local government have been similar in many ways. First, local authorities need to act as leaders in order to mobilise and reconcile varied interests within a community and thus adopt a more effective approach to tackle city pollution. Second, while becoming the coordinators or enablers of service delivery and the facilitators for public participation within an urban centre, they also need to be seen as legitimate (i.e. elected) actors within the whole system of governance. Third, in seeking an effective and coordinated response to urban environmental problems they need to enhance their capacity to work in partnership with other public and private sectors of the community. Finally, while working in public/private partnerships they need to build a network to allow financial and technical support from other local authorities and from the central government, local associations and non-governmental and private agencies (see, for example, Gilbert et al 1996; Hardoy et al 1992; Haughton and Hunter 1994). As it will be seen in following chapters, some of these issues are central to the discussion on reorganising the structures of local government in London and Mexico City in order to deal more effectively with urban environmental issues. In the case of London, concerns have been raised about the need for a more efficient, legitimate, and coordinated governmental response to air pollution at local level. In the case of Mexico City, while there are concerns over improving the structures of local government, a more effective response to air pollution seems to be constantly linked to enhancing the democratic local units of government. Before turning in more detail to the research on the case studies, this chapter will now consider the issue of urban air pollution.

Urban centres and air pollution: the scale of the problem

The significance of urban centres at the front of the environmental debate has been increasingly recognised because of the effects (positive and negative) they have on the natural environment (see Breheny 1992b: 277; Haughton and Hunter 1994: 1; Stren 1992a: 1). On the one hand, cities are key elements in the development of local, national and international economies. They are regarded as centres of production, exchange and consumption where cultural, artistic and social activities are manifested. On the other hand, cities are also seen as obsolete centres in the approaching age of advance information technology where a high degree of environmental contamination is produced (CEE 1990: 20-21; Elkin et al 1991: 4; OECD 1990: 9; Sherlock 1991: 13).

Table 2.1 Common environmental problems in urban centres: cause-effect relationship

	Source	*Effect*
Air pollution	Mainly from traffic, industry and waste incineration	Affects human health (respiratory illnesses; may contribute to death). Reduces human production potential and industry productivity. Affects atmospheric environment
Water pollution	Mainly from domestic and industrial sources (heavy industry and waste disposal activities)	Considerable risks for human health (deterioration in quality for human consumption i.e. drinking water). Harmful consequences to aquatic environment and industrial needs
Noise pollution	Mainly from road traffic, neighbourhood and aircraft noise	Human health is affected: disturbs sleep and causes stress
Waste	Mainly from household use and from radioactive, clinical and other 'hazardous' waste (hospitals, factories, industrial premises)	Dangerous to human health. It can contaminate air, water or soil
Contaminated land	From a variety of human activities: urban sprawl (conversion of land to urban uses); industry (metal extraction, mineral processing), industrial and domestic waste	Affects human health (particles of soil with irritant or poisonous chemicals) and agricultural uses

Source: Adapted from Murley (1994); OECD (1990); UNEP (1993); WRI (1992)

From an environmental perspective, urban centres play a complicated but essential ambivalent role (Stren 1992a: 1). While they are the major consumers of resources and the major producers of pollution and waste, they can also make a fundamental contribution to the solution of local, regional and global environmental problems (Breheny 1992a: 2; Gilbert et al 1996: 15; Gossop and Webb 1993: 129). As already seen, the environmental challenges for urban centres rest on the issues of resource-use and pollution reduction while achieving sustainable patterns of living. While urban centres continue to expand - by the year 2000 half of the world's population will be living in cities[3] - it is their environmental impact and not their size or wealth which has become the central issue in the move towards global and urban sustainability (see Gilbert et al 1996: 14-15; Haughton and Hunter 1994: 10). Certainly, although the degree of environmental deterioration varies within different cities around the world, city dwellers are exposed to a combined number of environmental problems whose damaging effects are not restricted to any particular size, age, or type of city (OECD 1990: 21-22). As seen in Table 2.1, common urban threats that can damage human health, flora and fauna and the built environment, include air, noise and water pollution, waste disposal, and derelict land. In particular, the adverse effects on the health and welfare of human beings and the environment caused by urban air pollution vary according to diverse existing air pollutants in urban centres - such negative effects are not only local, but regional (such as acid rain) and global (such as ozone depletion and global warming) (Haughton and Hunter 1994: 157). As seen in Table 2.2, the most common sources of these urban air pollutants are industry, production and consumption of energy and, above all, motor vehicles.

Achieving and maintaining healthy air quality in urban centres requires the adoption of an appropriate and effective urban air quality management regime. Ideally, as an integrated functional system, the latter should include such components or subfunctions as air quality monitoring networks, emission inventories, numerical prediction models, air quality standards and public information bands, and a range of cost-effective pollution control policies (e.g. public transport and traffic management measures) together with the resources and powers to impose them (see Elsom 1996: 67-68; Richardson 1992: 148-149). In establishing and operating an effective pollution control system for managing air quality, the question of what constitutes pollution to the atmosphere, therefore becomes essential. The lenses through which the concept of pollution has been seen and defined, and thus applied to air emissions, has considerably changed over the years. It has been usually said that pollution exists when enough of a substance or form of energy introduced by humans is present in the environment to produce harmful effects on it (see, for example, Blowers 1993c: 72; NSCA 1991: 133; RCEP 1971: 4). The contamination of the environment and its subsequent damage occurs when pollutants and sinks are not in balance, that is to say, when due to the nature or quantity of the substances released they cannot be readily absorbed by the ecosystem (see Button 1988; Haughton and Hunter 1994: 125; Levinson and Shetty 1992: 16; Strauss and Mainwaring 1984: 3). Nevertheless, there is no consensus on how much of these substances is 'enough' to pollute and cause harm to the environment. Hence, as argued by Ball and Bell (1994: 90) pollution is a relative concept because 'there is no absolute rule about what amounts to pollution'. It follows that air pollution exists only when the discharge of substances into the environment has been associated with damage or threat to humans and their health, other living species and their interrelationships, and to the natural and built environments (see, for example, Brañes 1994: 435; Elsom 1992: 3; EPA 1990; Strauss and Mainwaring 1984: 6).

The debate on defining air pollution moves then on to decide what level of pollution is permissible, taking two major considerations into account. First, the level established is usually the result of social, and above all, political constructions. In other words, the ultimate decision on what amounts to 'permissible levels of pollution' in a particular location (city, town, village) or for an entire nation or region, is social and particularly political. Thus, special attention needs to be paid to the diverse ways in which society and government perceive and look at pollution, particularly when doing research on a specific country or city from a comparative perspective. It may be mentioned, however, that standards and guidelines set up by international organisations - such as the World Health Organisation (WHO) - also play an important role in defining threshold limits of domestic and/or local pollution levels. Second, when determining these permissible levels of pollution, there must exist a trade-off between pollution itself and other factors such as society, political aims, model of economic development, financial costs, environmental quality standards, international legally-binding agreements, and so forth. The following sections examine the scale of the problem (origins and consequences) of air pollution in two different urban locations: London and Mexico City.

Table 2.2 Common urban air pollutants: cause and effect on human health

Pollutant	*Cause*	*Effect*
Asbestos	Heating and insulation materials, car clutch and brake linings	Pulmonary fibrosis (asbestosis). Pleural plaque, cancer of the lungs and cancer caused by asbestos
Carbon Monoxide	Incomplete combustion of carbon-based fuels: exhaust of petrol engines, industrial furnaces, power stations, faulty domestic cooking and heating appliances	Suffocation, central nervous system affected, headache with reduced mental activity-throbbing headache, vomiting and collapse-coma-death
Lead	Motor car exhaust accounting for about 75% (exists in the form of dust or fumes)	Central nervous systems affected (young people: foetuses), concentration and intelligence affected (especially in children even at low levels of exposure)
Nitrogen Oxides	Significant ones: nitric oxide and nitrogen dioxide. Combustion of fossil fuel (power generation, heating plants, road), industrial non-combustion processes	Nitrogen oxide: causes throat and eye irritation; respiratory illness in children. Nitric oxide: contributes to inhibit capacity of blood to carry oxygen round the body

Source: Murley (1994)

Table 2.2 Common urban air pollutants: cause and effect on human health (cont.)

Pollutant	*Cause*	*Effect*
Photochemical Oxidants	In the presence of sunlight nitrogen oxides react with hydrocarbons from vehicle emissions and industrial sources to produce ozone (a secondary pollutant)	Eye, nose and throat irritation, chest discomfort, cough and headache. Dangerous to people exercising or suffering bronchitis or asthma: reduces resistance of the lungs to disease
Sulphur Dioxide	Burning coal and oil	Irritant to eyes, mucous membranes, bronchio-constriction, stimulate coughing (dangerous in patients with cardio-respiratory problems)
Suspended Particulate Matter	Combustion products, industrial activities, natural sources (e.g. smoke)	Affects lungs [illegible] the air sacs, disturbance to the tissue
Volatile Organic Compounds	Exhaust fumes, cigarette smoke, synthetic materials and household chemicals	Toxic and carcinogenic chemicals. Anemia, eye, skin, throat irritation, nausea, allergic reactions, lung disease

Source: Murley (1994)

The case of London

Air pollution in London has a long history (see, for example, Ashby and Anderson 1981; Ball and Bell 1994; Brimblecombe 1987; Elsom 1992; RCEP 1971; 1974; 1976 and 1984). Although the earliest registered air pollution incident in England occurred in Nottingham, London is one of those few cities in which polluted air was detected and contested as early as 1273 (Murley 1994: 47). At that time, coal began to substitute wood as an industrial fuel and since then the usage of industrial coal (lime production) represented the main source of air pollution within medieval London.

Although pollution from coal burning was regarded as a problem in medieval times (Brimblecombe 1987: 10-11) air pollution in London grew more severe when the Industrial Revolution began. The rapid technological advance, demand on fuel consumption, the use of coal for industrial and domestic purposes, together with urban expansion in a growing industrial country during the seventeenth and eighteenth centuries started to have an important effect on London's atmosphere and consequently, on humans, flora and fauna, and the built environment (Brimblecombe 1987: 9-36, 65-68; Ball and Bell 1994: 286-303).

By the time the first Alkali Act was passed in 1863, London had already been experiencing high levels of polluted air - known as 'London smogs' or 'pea-soupers'[4] - which became more frequent and severe during the late Victorian times producing tragic effects. As seen in Table 2.3, during this period, and well into the mid-twentieth century, people died in London as a result of severe pollution episodes.[5] The worst took place on 5-8 December 1952 when a heavy concentration of smog (high pollution concentrations of sulphur dioxide and smoke) hung over the city with calm conditions (i.e. still air) and temperatures below freezing. This led to an increased number of deaths - nearly 4 000 - especially among the elderly, due to bronchitis, influenza, pneumonia, tuberculosis and other respiratory illnesses (Ball and Bell 1994: 288; Elsom 1992: 242-243). The number of health studies - mainly epidemiological investigations - on mortality and morbidity due to the infamous 'pea-soupers' in London, is now very well documented (see, for example, Brimblecombe 1987; DoH 1991; 1992; 1993; 1995 and 1997; Medina and Quenel 1993; Schwartz and Marcus 1990).

Table 2.3 Major London air pollution episodes and excess deaths (1873-1994)

Year	*Month*	*Duration (days)*	*Excess deaths*
1873	December	3	270-700
1880	January	4	700-1 100
1882	February	-	-
1891	December	-	-
1892	December	3	~1 000
1948	November	6	~300
1952	December	5	4 000
1956	January	-	480-1 000
1957	December	-	700-800
1959	-	-	200-250
1962	December	4	340-700
1975	December	3	-
1976	June	-	-
1982	November	-	-
1989	June	2	-
1991	December	4	100-180
1993	November	2	-
1994	December	1	-

Source: Brimblecombe (1987); DoH (1995 and 1997)

The levels of smoke, grit, dust and SO_2 in London during the 1950s and following years started to decline due to a number of independent factors acting at the same time. These included the initiatives of local authorities bringing in their own measures to control smoke and the efforts of the alkali inspectors. In addition, technological measures, such as the change-over to gas and electricity and the spread of central heating were also implemented. From a social point of view, migration from the city to the suburbs during those years also influenced air quality in London (see Ashby and Anderson 1981: 116-119).

The nature of air pollution in London over the last three decades is therefore substantially different from that of the past. Current concerns about polluted air in London are not related to the once feared 'London smogs' of the nineteenth and mid-twentieth centuries. Rather, they are related to the rising levels of contemporary or modern pollutants[6] associated primarily with road transport emissions (Bell 1993: 12-13;

Elsom 1996: 185; FoE 1994; QUARG 1993: 1; SEIPH 1994: I/2-I/3; Weir 1993: 1-3). In Britain, the 'traditional' pollutants have been identified as sulphur dioxide (SO_2), smoke, and other particulates that arise mainly from the combustion of coal or heavy oil for heating or power generation purposes. The 'newer' pollutants have been usually associated with motor vehicle emissions, although some of them also derive from heating or power generation sources, for example, nitrogen dioxide (NO_2) (see DoH 1993: 3). These contemporary and also common pollutants in other urban centres include carbon monoxide (CO), hydrocarbons (HC) (generally more volatile organic compounds (VOCs) such as benzene and 1,3-butadiene), nitrogen oxides (NO_X), and ozone (O_3). Suspended particulate matter (SPM) (especially black smoke and particles less than 10μm PM_{10}) and carbon dioxide (CO_2) have also been included within such a category (see, for example, WHO/UNEP 1992: 7-10).

Other sources of pollution in contemporary London include processes that also involve the combustion of fossil fuels; for instance, domestic and commercial building heating systems, plants producing heat and/or electric power for industry, and the two power stations which generate electricity for London Transport (LRC 1993: 119). Geographical and meteorological conditions have also played an important role in the formation of some pollutants such as nitrogen oxides or ozone. Depending on the temperature or the season these pollutants could be dominant when a pollution episode occurs (DoH 1992: 3-6 and 1993: 3-4; QUARG 1993: 29-30, 111-112).

The research that has been recently carried out by government-led and autonomous organisations, such as the Advisory Group, the Quality of Urban Air Review Group (QUARG), the London Research Centre (LRC), or the London Air Quality Network (LAQN), on the newer pollutants and their environmental effects, confirms the impact of vehicle emissions on contemporary London's air quality (see Tables 2.4 and 2.5). As seen in Table 2.6, car ownership in London has considerably increased over recent years, particularly due to many households acquiring a second or third car. According to the Department of Transport (DoT), growth in the number of cars available to London households is estimated at an average of 32% between 1991 and 2011. Nevertheless, the DoT has emphasised that car ownership forecasts are less relevant than car use (see DoT 1996: 72-73).

Table 2.4 Total emissions in London (1991) by sector (%)

Sector	CO_2 (as carbon)	SO_2	Black Smoke	CO	NO_X	VOCs
Road Transport	33.0	22.0	96.0	99.0	76.0	97.0
Other Transport	3.0	1.0	2.0	1.0	4.0	1.0
Electricity Supply Industry	2.0	0.0	0.0	0.0	1.0	0.0
Other Industry	13.0	43.0	1.0	0.0	5.0	1.0
Domestic	30.0	1.0	0.0	0.0	6.0	1.0
Other	19.0	32.0	2.0	0.0	8.0	0.0

Source: LRC (1993)

Table 2.5 Transport contribution to air pollution in London (%)

Pollutant	Road Traffic	Other Transport	Total Transport
Sulphur dioxide	22.0	1.0	23.0
Black smoke	96.0	1.0	97.0
Carbon monoxide	99.0	1.0	100.0
Nitrogen oxides	76.0	4.0	80.0
Hydrocarbons	97.0	1.0	98.0

Source: LRC (1993)

Table 2.6 Vehicle growth in London (available to households 1971-1991)

Area	1971	1981	1991	1971-1991 (% change)
Inner	472 000	502 000	580 000	23.0
Outer	1 392 000	1 682 000	1 969 000	41.0
All London	1 893 000	2 213 000	2 581 000	36.0

Source: DoT (1996)

In its Eighteenth Report (1994), the Royal Commission on Environmental Pollution (RCEP) emphasised its concerns that pollution caused by vehicle emissions may be adversely affecting human health in London. Nevertheless, the epidemiology of the health effects of the more recent air pollutants and of their mixtures in London's air, that is to say, basically those pollutants that are related to vehicle emissions, is still poorly developed (see DoH 1995: 84; SEIPH 1994: 3/13/7). The existing health reports relating to the latter,[7] have focused particularly on the respiratory effects of exposure to NO_2, O_3 and particulates. The recent concerns about the risk of cancer from exposure to benzene and other VOCs have also been included in such reports (see RCEP 1994: 28-31). The outcome of these studies shows not only that London can still experience air pollution episodes, but that there is growing evidence of the dangers that high levels of pollution may pose to human health (see Bell 1993: 18-25; DoH 1991: 67-87; 1992: 101-127; 1993: 89-116 and 1995: 83-118; SEIPH 1994: 3/1-3/7).

From a worldwide comparative perspective, London does not have the alarming levels of air pollution found in other urban centres. In a recent report of urban air pollution in twenty megacities of the world, jointly produced by WHO and UNEP (see WHO/UNEP 1992), concentrations of diverse air pollutants in London do not seem to regularly exceed the WHO health guidelines. For example, while the WHO guidelines are normally met in London for levels of O_3, these guidelines are normally exceeded in cities such as Los Angeles, Mexico City, Sao Paulo and Tokyo. Likewise, while London has low pollution levels of particulate matter (SPM), this pollutant is a serious problem for cities such as Bangkok, Beijing, Bombay, Cairo, Calcutta, Delhi, Jakarta, Karachi, Manila, Mexico City, Seoul and Shanghai (see WHO/UNEP 1992).

Within the European context, London is neither among the most polluted European cities nor among the cleanest. For instance, the levels of exposure for O_3 in 1991 ($\mu g/m^3$/day over 1 year) in London were lower than in Athens, Barcelona, Berlin, Lisbon and Turin. Likewise, over the same period, levels of exposure for SO_2 in London were lower than in Athens, Barcelona, Berlin and Turin, but were higher than in Lisbon, Paris and Warsaw. The levels of exposure for NO_2 were also higher in Athens, Madrid and Turin, but lower in Lisbon and Paris when compared to those in London (see Medina and Quenel 1993).

Nevertheless, among diverse UK cities, London seems to be one of the most polluted urban centres. For example, NO_2 concentrations over large parts of the Greater London area have been higher than in the rest of the UK and have exceeded the European Community (EC) Guide Value (Bell 1993: 15). Although cities such as Glasgow and Manchester have also experienced high concentrations of NO_2 in recent years, these have barely exceeded the EC Guide Value for NO_2 and have breached the WHO health guideline only a few times (see QUARG 1993: 48). Most significantly, the percentage of total emission of pollutants from road transport - such as CO, HC, black smoke, NO_X and SO_2 - is higher in London than across the entire UK (see SEIPH 1994: 4/2-4/3).

Improving air quality in Britain during the 1970s and 1980s does not seem to have been a government priority as the rising levels of new emerging pollutants in London - other than smoke and sulphur dioxide - were not promptly, or adequately, dealt with (Elsom 1992: 276). The reasons for this are threefold. First, there was lack of interest from the central government. Indeed, the slow and piecemeal government response to controlling air pollution during these years is closely related to the fact that many contemporary environmental affairs became part of the British public policy agenda only in the late 1980s (McCormick 1991). The belated recognition by the British government of the impact of acid rain[8] is an example of this. Despite the fact that many research groups and environmental organisations had been highlighting likely acid deposition effects for many years, it was not until the mid 1980s that the British government recognised the effects of acid rain on aquatic ecosystems and forests in the UK (Elsom 1992: 257). Hence, there was no reason for the British government to tackle such pollutants as NO_X and photochemical oxidants on the grounds of 'acid rain' damage.

Second, with the exception of lead,[9] pollution control acts and regulations continued to focus mainly on smoke and sulphur dioxide emissions from industrial and domestic sources (Elsom 1992: 244-250). Despite the increase in the levels of some pollutants in London from mobile sources, such as CO and NO_X during the 1980s (QUARG 1993: 32; 82-83), the government did not introduce new legislation until the EC started issuing Directives specifically requiring air quality standards. For instance, it was not until 1989 that the government formally introduced British legislation in this area via the Air Quality Standards Regulations which set the limit and guide values for SO_2, suspended particles, lead in air, and NO_2 previously set by EC Directives (DoH 1993: 10; QUARG 1993: 174). A considerable part of the current British environmental policy and body of legislation regarding air quality has actually been formed mainly due to EC directives, regulations and other measurements (see Ball and Bell 1994: 70; McCormick 1991: 20; O'Riordan and Weale 1990: 2).

It must be noted that the UK, since joining the EC in 1973, has consistently displayed resistance and delay with regard to the implementation of EC Environmental Directives. For example, in 1980 the EC set out the maximum concentrations of smoke and SO_2 permitted in urban areas through a Directive (80/779/EEC). This Directive, however, was not implemented for four years, the reason being that the adoption of the EC strategy on air quality standards was different to the then 'best practicable means (BPM)' approach to pollution control adopted in the UK. The UK approach relied on voluntary compliance by the polluters (industrialists) with unclear and non-mandatory guidelines or reference levels, rather than more clearly defined and mandatory standards (Elsom 1992: 260; 274). The 'new' approach adopted in Part I of the EPA in 1990, known as 'Best Available Techniques Not Entailing Excessive Cost (BATNEEC)', embraces a similar non-mandatory system and, thus, seems to suggest further delay in the implementation of EU Directives on the part of the UK government (see Ball and Bell 1994: 21-22; Elsom 1992: 240-241).

Finally, although concern about vehicle emissions emerged in the 1970s, there was no evidence in Britain that such emissions would seriously damage human health (Ashby and Anderson 1981: 143). This, however, led to a considerable coverage on specific health aspects, such as asthma. Again, without reliable and sufficient data on levels of emerging pollutants and their impact on human health, policies centred on stationary sources. It was not until the 1990s, particularly with the Royal Commission on Environmental Pollution RCEP's Eighteenth Report, that the link between air pollution and transport became more pronounced (RCEP 1994). While during the 1980s and 1990s media and increasing public concern put some pressure on the British government to respond to air pollution, it was not until the mid-1990s that the latter identified urban air quality as a priority area for improvement within its 1994 sustainable development strategy (see Ball and Bell 1995: 327; Bell 1993: 7; DoE 1994: 49-57; Elsom 1996: 185).

The need to improve London's air quality has recently led the British government to develop diverse traffic management and public transport policies as well as to draw up a new air quality strategy by the mid-1990s. For instance, in 1995, variable speed limits were introduced on London's orbital motorway - the M25 - to smooth traffic flow and reduce congestion in order to decrease the number of stop-start occasions when vehicles are moving slowly or are stationary with engines idling and creating higher emissions. Another example is the creation of the Red Routes system throughout London, which consists of designating urban clearways where stopping is banned during working hours. This measure (initiated in 1991 with a pilot scheme in north and east London) aims at shortening car and bus journey times so as to ease traffic congestion (see DETR 1997a: 7; Elsom 1996: 188-189). Early in 1995, a new air quality strategy was outlined in a Consultation Paper called 'Air Quality: Meeting the Challenge' (see DoE 1995). Later that year, the 1995 Environment Act (EA), laid down the principles of a coherent air quality management strategy and introduced the UK National Air Quality Strategy (NAQS) which was adopted by the beginning of 1997 (see DoE 1997). The 1995 Environment Act and the 1997 NAQS provide not only for the further development of local air quality assessment and management, but also new regulatory powers for the improvement of air quality giving London authorities the powers they need to tackle London's episodic and long term air quality problems. Fulfilment of the specified air quality standards targets, however, has been put off until 2005, allegedly, to allow EU directives to come into force and to implement other measures such as adapting refined fuel to new vehicles. According to the central government, such delayed targets are 'reasonable and justifiable on consideration of the costs and benefits' according to the standards and objectives laid down in the strategy (DoE 1996: 17; Elsom 1996: 192-193).

Undoubtedly, over the last 30 years, under-investment in public transport, the UK government's long-standing commitment to road building, and the growth of car ownership and car use, have largely influenced air quality in London. For instance, despite an increase of government spending in the London Underground system over recent years,[10] the road building programmes - which have included proposals to widen London's orbital motorway - have remained an investment priority. Indeed, road transport is an area where not has only public expenditure risen sharply - more than 50% for 1994/95 in real terms compared with the level in the 1980s - but where there exists the only long-term investment programme within the government policy framework (Elsom 1996: 14 and 191; RCEP 1994: 82). Hence, further investment in public transportation systems in London has been continuously called for from the central government (see, for example, Bell 1993: 38; Dobson 1995?: 11; LRC 1993: 114-115; RCEP 1994: 15-16 and 244).

The case of Mexico City

Environmental transformation in the Valley of Mexico arguably started approximately six centuries ago. According to recent sources, during Pre-hispanic times, the capital of the Aztec *Señorío* Tenochtitlan (1324-1521), could not sustain itself with the existing natural resources of the valley, mainly due to its population growth (aprox. 300 000 in. by 1500) and had to import maize, beans, tropical fruits, salt, wood, and so on from other regions (CMPCCAVM 1994c: 9; Gamboa de Buen 1994: 19-22). It was not until the Colonial (1521-1810) followed by the Independent (1810-1910) and the Revolutionary (1911-up to present days) periods, though, that environmental degradation of the valley began: frequent droughts due to the new city-planning; desiccation of surrounding lakes; soil erosion; forest devastation; shortages of water supply and contamination; and, more recently, high levels of air pollution (Ezcurra 1995; Fried 1972: 647-654; Schteingart 1989: 44).

The existence of air pollution was detected and contested in Mexico City during the 1950s and 1960s. The origins of air pollution in Mexico City derive from rapid industrial development growth together with urban and population expansion which have resulted in a permanent and intense energy (fuel) consumption for maintaining industrial productivity, electricity generation, public services, household amenities, transportation, and so on (Bravo Alvarez 1987: 127; CMPCCAVM 1994b: 4; Collins and Scott 1993: 120; Díaz Díaz and Perló Cohen 1994: 44; Pezzoli 1991: 205-207). The particular geographical location and meteorological conditions of the city[11] have also played a major role in the formation of certain pollutants, such as O_3.

Emission of pollutants in Mexico has been usually associated with three types of sources: fixed, moving and natural sources. Although Mexican legislation has classified and defined them in different ways (see LFPA 1982; LGEEPA 1988), fixed sources are commonly identified with all types of industry or any combustion process that generates pollution; moving sources refer to all types of motor vehicles; and natural sources are associated with *'tolvaneras'* which encompass dust and soil particles (see Bravo Alvarez 1987: 136-164; CMPCCAVM 1994a: I/4; Gamboa de Buen 1994: 139). The 1989 Emissions Inventory for Mexico City identified emission sources by sector dividing them into energy, industry and services, transport, and environmental damage. As shown in Tables 2.7 and 2.8, according to governmental reports, moving sources - i.e. transport - appeared by the late 1980s as the main source of pollution in Mexico City by weight (76.7%) and by toxicity (42.4%) as opposed to natural environmental degradation (15.0% by weight and 29.9% by toxicity) and fixed sources - energy, industry and services (8.4% by weight and 27.7% by toxicity).

Table 2.7 Emissions Inventory (1989) by sector (% by weight)

	SO_2	NO_X	HC	CO	SPM	*Total*
Sector						
Energy	35.5	5.6	5.6	1.8	1.0	4.0
Industry and Services	42.7	18.5	7.0	0.6	2.8	4.4
Transport	21.8	75.4	52.5	96.7	2.1	76.7
Ecological Degradation	0.1	0.5	34.9	0.9	94.0	15.0
Total	100.0	100.0	100.0	100.0	100.0	100.0

Source: STI (1990)

Table 2.8 Emissions Inventory (1989) by sector (% by toxicity)

	SO_2	NO_X	HC	CO	SPM	*Total*
Sector						
Energy	7.2	1.1	1.3	0.2	1.0	10.8
Industry and Services	8.6	3.7	1.7	0.0	2.8	16.9
Transport	4.4	14.9	12.6	8.4	2.1	42.4
Ecological Degradation	0.0	0.1	8.4	0.1	21.3	29.9
Total	20.2	19.8	23.9	8.7	27.3	100.0

Source: STI (1990)

During the late 1980s, the levels of traditional urban pollutants, such as SO_2, SPM, Pb and NO_X, were all well above, with the exception of CO, the WHO health guidelines (see Calderón-Garcidueñas et al 1992: 225; Kandell 1988: 528; Schteingart 1989: 44; WHO/UNEP 1992: 155-163). By the late 1980s and early 1990s, the high levels of air pollution - among other environmental problems - gave Mexico City the status of one of the most polluted cities in the world:

> Mexico City, recently referred to as the 'anteroom to an ecological Hiroshima', has become one of the most contaminated, unhealthy cities in the world...some authors have declared it a 'negative urban ecosystem or antiecosystem' (Pezzoli 1991: 205).

Within a Latin American context, Mexico City is the most polluted city for several pollutants. In one of the most recent comparative reports (see WHO/UNEP 1992) the levels of air pollutants - such as SO_2, SPM and CO - in Mexico City were higher than in cities like Buenos Aires in Argentina and Rio de Janeiro or Sao Paulo in Brazil. The capital of Chile, Santiago de Chile, seems to parallel Mexico City's bad air quality situation for particulates as it has been found that particulate concentrations in Santiago de Chile are among the highest observed in any urban area in the world (Ostro et al 1995: 6). In the case of O_3 although this pollutant has become a serious problem for Mexico City and Sao Paulo, the ambient concentrations in Mexico City are exceptionally high compared to any other city (see Medina and Quenel 1993; WHO/UNEP 1992). Following Mexican government reports on levels of ozone, it has been pointed out that the Mexican Air Quality Norm for this pollutant (0.11 ppm / 1 hour) was breached 750 and 959 hours in 1987 and 1988 respectively, and from 1989 to the first half of the 1990s, more than 1 000 hours on average every year. This amazing figure derives from the fact that the threshold limits of the Mexican IMECA index value[12] are too high compared to international values (Bravo Alvarez and Torres Jardón 1995: 10; Calderón-Garcidueñas et al 1992: 225; Calvillo 1993: 5; Campos Ruíz et al 1993: 71; Collins and Scott 1993: 123-124; Quadri 1994: 24). Indeed, whereas the Mexican value norm for ozone is 100 IMECA points which means a concentration of 0.11 ppm or 216 $\mu g/m^3$ in one hour, the WHO standards are 150-200 $\mu g/m^3$ in one hour (time-weighted average). Although the latter may not represent a big difference within comparative worldwide guidelines (see Table 2.9), the contingency plan in Mexico

City only starts operating when ozone levels reach 250 IMECA points which means a concentration of 0.29 ppm in one hour (CMPCCAVM 1995b: VII/19; SEDESOL 1993: 166-167; WHO/UNEP 1992: 226).

Table 2.9 Guidelines for O_3 (ppm / 1 hour)

	WHO	EU	USA	MEXICO
Guidelines	0.05-0.10	0.076-0.10	0.12	0.11

Source: INEGI (1994a: 395)

Despite the existing poor air quality situation in Mexico City, there are very few studies that have reviewed the relation between air pollution and increased mortality rates. The first study of time series analysis carried out in the Metropolitan Zone of Mexico City (MZMC) for the period 1987-1989, showed a positive and significative autocorrelation between SPM and SO_2 with mortality (Santos-Burgoa and Rojas Bracho 1992: 229). A second study-analysis carried out by the Pan American Health Organisation (PAHO) and the Mexican Secretariat of Health (Ssa), estimated that the concentrations of pollutants in Mexico City could be producing an annual mortality rate of 5% for all pollutants. The latter could have meant that during the early 1990s, air pollution led to an increased number of 800 excess deaths in the Northwest area (Xalostoc) and 600 excess deaths in the Southwest region (Pedregal) annually (Calvillo 1993: 27). In another study, carried out for the World Bank in 1992, conservative calculations suggested that there was a significant relation between total suspended matter (TSP) and mortality rates. Based on the levels of TSP pollution in Mexico City during 1990, the estimated total number of (statistical) lives saved would have been 6 400 - equivalent to 3.8 lives per 10 000 people (Margulis 1992: 13).[13]

The number of health studies on morbidity in the MZMC surpasses by large those on mortality.[14] Over the last years, epidemiological studies have concentrated on respiratory illnesses and other human health effects during the occurrence of air pollution episodes or environmental contingencies in Mexico City. Most health studies, though, have focused on the analysis of the effects of lead, ozone, and more recently, particles (see, for example, Calderón-Garcidueñas et al 1992; Calvillo 1993; DDF

1996; Hernández Avila 1995?; Margulis 1992; Medina and Quenel 1993; Restrepo 1992; Ssa 1993). For instance, ozone studies have included the cases of school absenteeism among children and childhood asthma in Mexico City. In both cases the results have suggested that ozone exposure may have adverse effects on the respiratory health of children and that they are positively associated with the number of children's emergency visits for asthma in Mexico City (see Romieu et al 1993 and 1995?).

Table 2.10 Total atmospheric emissions (SO_2, NO_X, HC, CO, SPM) for the MZMC according to official figures (tons/yr)

Year	*Tons.*
1972	2 653
1973	2 868
1974	3 244
1975	3 394
1976	3 431
1977	3 438
1978	3 449
1979	3 521
1980	3 600
1981	3 672
1982	3 757
1983	3 851
1984-1987	-
1988	**4 900**
1989	4 356
1990	-
1991	4 300
1992-1993	-
1994	4 009
1995-1997	-

Source: CMPCCAVM (1992); DDF (1996); Schteingart (1989); STI (1990)

Undoubtedly, the high levels of air pollution that have been experienced in the MZMC has become one of the main preoccupations for all sectors of Mexican society (see Cancino Aguilar 1994: 105; CMPCCAVM 1992: 7; Díaz Díaz and Perló Cohen 1994: 43-44; Ezcurra

1990: 583 and 1995; SEDESOL 1993: 153). Despite a rapid increase of around 45% in total emissions to the atmosphere during the 1970s and 1980s (see Table 2.10) though, it was not until 1986 that polluted air became a priority for the Mexican government. Previous official attempts to control air pollution - from 1972 (when the *Subsecretaría de Mejoramiento del Ambiente* was created) to 1986 - were scarce and insufficient (see, for example, Schteingart 1989: 47-48). The latter can be explained partly because recognition and concern about environmental issues in the Mexican government did not come until the last years of the six-year mandate *'sexenio'* (1982-1988) of former President Miguel de la Madrid Hurtado. Before the years of 1987 and 1988, the government response to environmental issues was characterised for being heavily centralised, specific or 'sectorial', and non-democratic (Brañes 1994: 157; Nuccio et al 1993: 270).

During de la Madrid's administration in 1986, a series of plans were set out to overcome the increasing air pollution problem in the capital of the country. By way of illustration, some of the new proposals for the MZMC included an industry relocation programme, a vehicle emission control system, the establishment of an automatic air quality monitoring network, unleaded petrol for motor vehicles, and the change-over to natural gas in power plants (see, for example, Bravo Alvarez 1987: 237-255; Díaz Díaz and Perló Cohen 1994: 46; Gamboa de Buen 1994: 138). It was not until President Carlos Salinas de Gortari (1988-1994) came to power however, that a much stricter plan to control air pollution than that of his predecessor was launched in 1990 for the MZMC: the Comprehensive Program Against Air Pollution - *Programa Integral Contra la Contaminación Atmosférica de la Zona Metropolitana de la Ciudad de México* (PICCA). This programme, containing 41 measures,[15] focused on five main areas of concern: better quality of fuels, expansion of public transportation and strict vehicle emission control, improvement of combustion processes and emission control in industries and service establishments, environmental restoration, and environmental education and research as well as social participation. As a number of legal and economic measures took place, a new inter-governmental agency was also created: the Metropolitan Commission for Pollution Prevention and Control in the Valley of Mexico - *Comisión Metropolitana para la Prevención y Control de la Contaminación Ambiental en el Valle de México* (CMPCCAVM). This body - which changed its name in 1996 to the current Environmental Metropolitan Commission - *Comisión*

Ambiental Metropolitana (CAM) - has been since then in charge of coordinating air pollution prevention and control activities in the MZMC (see CMPCCAVM 1994a and 1994c; DDF 1996; SEDESOL 1993 and 1994; STI 1990). With such a responsive attitude, no one would have doubted that the Salinas administration had shown political will to overcome the problem:

> I am giving precise, urgent and imperative instructions to the Mayor of the Federal District to act immediately and efficiently to promote community participation in the fight against pollution[16] (STI 1990: 2).

The results that are being exhibited after a six-year period of dealing with polluted air in the MZMC though, raise the question whether the speech given by Salinas in 1988 represented only good intentions seeking short and immediate positive results for socio-political reasons,[17] rather than properly addressing and responding to the problem. Although serious governmental policy response only commenced in the late 1980s following the first detailed 1989 Emissions Inventory, and more properly, since 1990 through the PICCA, policy on air pollution control has already been contested (see Díaz Díaz and Perló Cohen 1994: 46; Gamboa de Buen 1994: 138-139; STI 1990: 31-33). Criticisms have focused on the fact that much of the latter had been formulated without following scientific recommendations and previous international experiences on combating some air pollutants - such as Pb and SO_2 - where control has resulted in unexpected side-effects. Indeed, the strategies for reducing Pb and SO_2 through changes on the content of petrol for motor vehicles and the change-over from fuel oil to natural gas in power plants during the late 1980s, respectively, have increased the emissions of HC and NO_X contributing to the formation of high levels of ozone (Bravo Alvarez and Torres Jardón 1995: 3-4; Ezcurra 1990: 583).

Although measurements of Pb are now well below international standards (1.5 micrograms per cubic metre) and NO_2, SO_2 and CO occasionally exceed the Mexican air quality norm, Mexico City still experiences heavy air pollution (see DDF 1996; SEDESOL 1994: 223-232). For instance, two of the main pollutants that were not a serious problem before the implementation of the 1986 and *a posteriori* regulations but are now constantly present in Mexico City's air, include suspended particles (such as PM_{10}) and O_3 (combination of HC + NO_X + solar light) (Bravo Alvarez and Torres Jardón 1995: 3; Campos Ruíz et al

1993: 73; CMPCCAVM 1994c 40-64; Collins and Scott 1993: 123-124; Hardie et al 1995).

Another criticism of the Mexican air pollution control system lies in the fact that the 1989 Emissions Inventory and subsequent official publications have mistakenly argued and informed on the sources of some pollutants, and consequently, inadequate policies have been adopted and implemented. For instance, the Mexican government has said that 95% of suspended particles are generated due to natural sources (environmental degradation or deforested areas) in the Basin of Mexico City. The latter led the government to elaborate a reforestation programme in 1990 which meant seeding 41.6 million trees in the urban and rural areas of the basin; 40% of these trees have already disappeared (CMPCCAVM 1994a: II/39 and 1994c: 60, 86). However, it has been argued that the primary sources of the emission of particles are combustion processes and fixed sources, and not *'tolvaneras'* (Bravo Alvarez and Torres Jardón 1995: 7). The 1994 Emissions Inventory shows that the percentage by weight of suspended particles from environmental degradation is still the same in comparison to the 1989 Emissions Inventory (see Tables 2.7 and 2.11). If the government is right in saying that the main source of particles is *'tolvaneras'* and that the strategy for bringing down their ambient concentrations is by seeding trees, then its approach has not worked.[18] Suspended particles in the MZMC are still above the Mexican air quality norm and have become the second biggest problem in Mexico City's atmosphere after ozone (Bravo Alvarez and Torres Jardón 1995: 7; Calderón-Garciduenas et al 1992: 225; Hardie et al 1995).

Table 2.11 Emissions Inventory (1994) by sector (% by weight)

	SPM	SO_2	CO	NO_X	HC	*Total*
Sector						
Industry	1.4	57.3	0.4	24.5	3.2	3.0
Services	0.2	15.9	0.1	4.2	38.9	10.0
Transport	4.2	26.8	99.5	71.3	54.1	75.0
Ecological Degradation	94.2	0.0	0.0	0.0	3.8	12.0
Total	100.0	100.0	100.0	100.0	100.0	100.0

Source: DDF (1996)

Most of the factors that have largely contributed to Mexico City's current air quality situation are still playing a major role within the policy-formulation process. For example, despite the governmental efforts to contain urban sprawl through land-use regulations and the creation of a green belt, urban and population growth continues to expand in the MZMC, including land zones for environmental conservation (CMPCCAVM 1994c: 85-86; Gamboa de Buen 1994: 123-138; Nava Escudero 1992: 114-116; Pezzoli 1991: 207). Although some industries in Mexico City have been closed - such as the *18 de Marzo* refinery - or have moved out of the city - such as General Motors - industrial relocation programmes have not succeeded and emission-control equipment has not been thoroughly installed due to high costs (see Collins and Scott 1993: 126). Government attempts to relocate industry by the end of the Salinas administration failed due to the economic crisis of 1994. Only a few large companies are financially prepared to move out of Mexico City once economic stability returns to Mexico.

By the time the Zedillo administration (1994-2000) came to power in the mid-1990s, it was clear that the air quality situation in Mexico City was still considerably poor. While previous pollution control measures did actually bring some improved air quality by the end of the Salinas administration, the high concentration levels of such pollutants as O_3 Particles and VOCs remain an unresolved problem. So, for example, although the peak levels of ozone have not exceeded more than 300 IMECA points over the last few years (this band was breached 11 days in 1992) and the number of times the contingency plan has been implemented has diminished from 7 in 1996 to 3 in 1997, the Mexican air quality norm - 100 IMECA points - is being breached annually more than 90% of the total number of days (see DDF 1996). The increasing concentration levels of diverse organic compounds in particular aldehydes which contribute to the formation of photochemical pollution - during the period 1993-1996, constitute another example of continuous poor air quality in Mexico City (see García et al 1998: 31). This situation led the Mexican government to launch a more integrated and even stricter air quality strategy in 1996: the Air Quality Improvement Programme - *Programa para Mejorar la Calidad del Aire en el Valle de Mexico* (Pro-Aire). This new strategy (which updates and enhances the previous air pollution programme PICCA), focuses on four main areas of concern: industry, private vehicles, public transport, and environmental restoration. In doing so, it seeks to diminish the number and concentrations of pollutants per day as well as to reduce the number of contingency or emergency situations per year.

Undoubtedly, the growth in car ownership and car use, the Mexican government's long- standing commitment to road building, and the lack of a more integrated and enhanced public transport system for the whole Metropolitan Zone of Mexico City - the MZMC - have largely influenced the air quality situation in Mexico City. Energy consumption continues to increase through rising levels of petrol, fuel oil, diesel and natural gas consumption (see Bauer and Quintanilla 1995). For instance, in the first four years of this decade daily petrol consumption in the MZMC increased 13.66% from 16.1 million litres every day in 1990 to 18.3 by late-1994 (CMPCCAVM 1995a: 9). This is partly explained due to the rapid growth of car ownership: the annual car sales in the entire MZMC grew from aprox. 100 000 units in 1983 to almost 250 000 in 1992 (see Table 2.12). Although these figures started to decrease for the years 1993-1995 (sales fell dramatically due to the 1994 economic crisis), car ownership and car use are expected to increase after financial recovery (see CMPCCAVM 1994c: 27; DDF 1996: 86).

Table 2.12 Vehicle growth and inhabitants/vehicles percentages in the Federal District (1940-1990)

	1940	1950	1960	1970	1980	1990
Vehicles	46 361	72 189	234 638	676 005	1 803 559	2 200 000 (aprox.)
Inhabitants per vehicle	37.1	42.6	20.7	19.1	4.9	4.0 (aprox.)

Source: Díaz Díaz and Perló Cohen (1994); *Excélsior* (June 1st, 1995)

Whereas the use of the private vehicle has been favoured through diverse road building programmes over the last few years (see Ward 1998: 145) the public transportation system in Mexico City has been insufficient, corrupt, and at times, subjected to significant financial cuts. So, for example, during the periods 1972-1976 and 1989-1990 the expansion and improvement of the underground system *Sistema de Transporte Colectivo* (STC) - *Metro* stopped. While the underground system has been systematically enlarged since the early 1990s, it has failed to discourage motorists from using their cars (see Navarro 1993: 41-55). Although the recent transportation plan for Mexico City - launched by the government in 1995 - contemplates the expansion and improvement of the public transportation system (e.g. underground, trolley buses and trams, buses) it will continue to favour private vehicles (see Chapter 7). This ambitious programme includes the creation and enlargement of dual carriage-ways and motorways within Mexico City and around the metropolitan zone: new ringroads are already under construction within and outside Mexico City (see Calvillo 1995; DDF 1995).

Conclusion

Within the context of the current environmental debate with regard to cities, air pollution represents an issue of significant urban environmental concern in countries of the North and the South. Diverse ways of approaching air quality management (both similarities and differences) within a North-South context greatly justify carrying out a comparative exercise on London and Mexico City. As seen in this chapter, the two cities that are the subject of the case studies both suffer from poor air quality - albeit the scale of the problem is different in each of them. While the levels of air pollution are far higher and more alarming in Mexico City than in London, the way to improve air quality seems to point in the same direction: tackling the use of the car and other road vehicles. The challenge for urban centres like London and Mexico City is to improve air quality without compromising the need to meet the demands for urban mobility. Measures such as traffic calming, vehicles emission control, banning of cars, or improvement of fuels, may help to improve the problem. However, as long as car ownership and car use continue to increase, traffic management policies and better public transport systems may be offset by increasing levels of road transport emissions. Thus, while short-term policies may be desirable for preventing present generations being exposed to high levels of pollution, medium to long-term policies are essential if future generations are to meet their own needs for air quality.

While an adequate and integrated response to achieving and maintaining healthy air quality in urban centres requires the participation of diverse actors in society, participants at local level play a crucial role in this. As seen in the following chapter, local authorities constitute the institutional means at local level for ensuring an efficient, coordinated and legitimate response for dealing with urban environmental issues, particularly regarding air quality management. The next chapter sets out the significant role that local authorities have for improving the environmental conditions of urban centres, and outlines the debates on models for reforming local government to improve air quality. The following chapters (Chapters 4-7) then describe and analyse how London and Mexico City's systems of local government operate in relation to air quality management.

Notes

1 The Cocoyoc Declaration - also known as 'Founex II' - was held in Cocoyoc, Mexico, in October 1974, in order to discuss the relationship between environment and development and to analyse the impact that environmental issues were having on development strategies and international economic relations (see McCormick 1995: 183-184).

2 Some efforts that have been made in the environmentalist thinking during the 1980s and 1990s to identify the variety of approaches and competing positions relating to sustainable development include Daly (1993), MacManus (1996), and Turner (1993).

3 Population rates in developed countries have nearly doubled since 1950, from 447 to 838 million, and in developing countries they have nearly quadrupled from 286 to 1.14 billion. In 1940, while 1 in 100 lived in a city of 1 million or more inhabitants, in 1980, 1 in 10 lived in a city of the same number of inhabitants (WCED 1987: 16, 236-237).

4 The so-called 'London smogs' were the result of 'smog' - a synthesis of fog and high levels of smoke, sulphur dioxide (SO_2), and meteorological conditions, for example, temperatures falling below freezing (see Elsom 1992: 22-23 and 242-244).

5 In Britain, the Advisory Group - which was set up by the Department of Health to provide advice to the Chief Medical Officer about personal protective measures during air pollution episodes - explained in its last report that an episode of elevated air pollution is to some extent arbitrary and that the criteria for it has changed over the years. This Group has identified three types of air pollution episodes in today's Britain. First, the 'summer smog' a pollution mixture with the main, or indicator, pollutant being ozone. Second, 'vehicle smog' where the indicator pollutant are nitrogen oxides. Third, 'winter smog' where the indicator pollutant is sulphur dioxide with a contribution from nitrogen oxides. For some further explanations on this including case studies of pollution episodes, see DoH (1995 and 1997).

6 There is not a standardised or universal classification for 'traditional' or 'contemporary' pollutants. For example, whereas for Britain modern pollutants

usually include CO, NO_X and O_3 (see DoH 1993: 3), for the OECD all of them are traditional pollutants (see OECD 1995: 13-14).

7 One of the most recent and significant bodies that has fostered the understanding of air pollutants and their human health effects is the Advisory Group. So far, this group has concentrated on O_3 (First Report); SO_2, acid aerosols and particulates (Second Report); NO_X (Third Report); and health effects from exposure to mixtures of air pollutants (Fourth Report) (see DoH 1991; 1992; 1993 and 1995). Another example is that of QUARG (Quality of Urban Air Review Group), a working group of experts established by the Department of the Environment to review current knowledge on urban air quality. Its First Report presented a fairly comprehensive description of nitrogen, organic and sulphur compounds, CO, particulate matter, oxidants, and metals (see QUARG 1993). Two other significant studies are worth mentioning. First, a report prepared by the London Research Centre with support of the European Commission. This report focused on a study of energy use in Greater London and the opportunities for improving efficiency and reducing pollution (see LRC 1993).). Second, the air quality reports of the London Air Quality Network (LAQN) produced by the South East Institute of Public Health (SEIPH) in conjunction with the Association of London Authorities (ALA) and the London Boroughs Association (LBA). These report present air quality information across London in terms of long term trends and peaks, and for the first time, by local authority area. The results include NO_2, CO, O_3, SO_2, particles (PM_{10}), hydrocarbons (benzene), and smoke (see SEIPH: 1994).

8 The main contributors to the creation of acid rain are SO_2, NO_X and photochemical pollutants (see Elsom 1992: 310-317).

9 As explained by Elsom (1992: 263), lead in petrol became a national issue in the United Kingdom in 1982 due to the lobbying of a pressure group called Campaign for Lead-Free Air (CLEAR). British action on phasing out leaded petrol started before this year but increased with this public campaigns and further EC Directives.

10 Since the mid-1980s, investment in the London Underground system increased from around £250 million a year to a stable level of a little over £500 million a year over the last three years. Although London Underground currently accounts for the largest share of investment in transport in London, there have been some significant cuts during the 1990s as opposed to an untouched road building spending programme: whereas investment for London Transport was cut by one-third in 1992, the roads programme remained intact. In 1993, the UK government announced that the M25 was to be widened to 14 lanes along its busiest road; the following year, in 1994, another cut to London Underground core investment was implemented. Before the general election of May 1997, the then DoT announced that other parts of the M25 would be widened within a longer term scheme (see Bell 1993: 38; DoT 1996: 63-64; 90; 212; Elsom 1996: 188).

11 The climate conditions and the geographical location of the Basin of Mexico contribute enormously to the formation or dispersion of certain pollutants. For example, due to high surrounding mountain chains on all sides except the North, wind speed and ventilation are limited, constraining the dispersion of pollutants. The predominant wind pattern, from North to South, carries pollutants emitted by heavy

industry, urban and transportation activities localised in the most concentrated areas to the rest of the urban conglomerate, i.e. Northern and Central parts. Additionally, the atmosphere is badly affected by the frequent 'thermal inversion' phenomena which is produced when a mass of cold air, at a certain height, impedes the renewal of warmer air making the pollutants stagnate at surface level during the early morning hours until the inversion is broken at midday. Whereas during the rainy season (May-October) the inversion is upset when a cold air mass penetrates, during the dry season (November-April) the inversion is more frequent and lasts several hours. The high solar radiation experienced in the basin also intensifies photochemical reactions between NO_X and HC which when all three combined, favour the formation of O_3. Another important factor is that due to its altitude (2 244 metres above sea level) oxygen content in the atmosphere of the basin is 23% less than at sea level causing, as a consequence, all processes of combustion to be less efficient. Fuel combustion by vehicles is certainly affected by this situation. For a more detailed description of the latter, see Bravo Alvarez (1987: 127-136); Bravo Alvarez and Torres Jardón (1995: 2-3); Collins and Scott (1993: 120-122); CMPCCAVM (1992: 7-13; 1994a: I/5; 1994b: 5 and 1994c: 16-19); Fried (1972: 647); Schteingart (1989: 44); STI (1990: 13).

12 The IMECA *(Indice Metropolitano de la Calidad del Aire)* is the Mexican index value for air quality measurement in the MZMC. The IMECA calculation makes the criterion value (either ppm or $\mu g/m^3$) for each pollutant equal to 100 points. For example IMECA 100 means 0.11 ppm in one hour for O_3; 0.13 ppm in 24 hours for SO_2; 13 ppm in 8 hours for CO; 0.21 ppm in one hour for NO_2; and 275 $\mu g/m^3$ in 24 hours for TSM. For a good explanation of the IMECA index value and its conversion for most pollutants for the MZMC, see SEDESOL (1993: 155 and 1994: 217); STI (1990: 29); WHO/UNEP (1992: 157-158).

13 It is important to note that the three cases quoted here contain a lot of methodological limitations which are an obvious result of the kind of analysis that is being used.

14 Some of the most important non government-led organisations that have carried out studies on air pollution and health effects in the MZMC include the Pan American Health Organisation (PAHO), Hospital ABC, and the Centre of Public Health Research (see Hernández Avila 1995?). The central government Secretariat of Health has also participated and coordinated some research on air pollutants and their effects in Mexico City through the *'Sistema Nacional de Salud'* (see, for example, Ssa 1993).

15 For a detailed description of the 41 measures see STI (1990: 34-41).

16 Ex-President Salinas de Gortari's inauguration speech in December 1988. Quote taken from WHO/UNEP (1992: 163); the original quote can be seen in STI (1990: 2): *'...doy instrucciones precisas, urgentes y enérgicas al Jefe del Departamento del Distrito Federal para que actúe de inmediato con acciones eficaces alentando la participación de la comunidad para abatir la contaminación'*.

17 The 1988 Presidential elections were the most contested ever in Revolutionary Mexico's history and have been regarded as fraudulent: Mr. Salinas' popularity and political legitimacy were well below all of his predecessors' (Ward 1990: 66-67).

18 Apparently, some of the information provided for the 1994 Emissions Inventory relating to particulates was taken from studies done in 1990. If this is the case, then the Mexican government has failed to update the Emissions Inventory and the follow-up to PICCA seems difficult to achieve (see DDF 1996). Other figures which are hard to believe are those relating to HC. While environmental degradation contributed to 34.9% of HC in 1989, in less than 6 years, by 1994, it decreased to 3.8% (see Tables 2.7 and 2.11).

3 Models of Local Government

Obviously times and conditions change and governmental institutions are always likely to change to match societal changes. However, it would be a shade premature to write off the metro idea just yet.

L.J. Sharpe (1995)
on the future of metropolitan government

Introduction

The need to reorganise local government in countries of the North stemmed primarily from the considerable growth of urban centres beyond the existing boundaries of their local governments, and from the expansion of local government's service functions, especially during the post-war period. Reviewing the organisation of local government, thus, has implied improving government structures in metropolitan areas where the enlargement or scale of local units has been at the core of the debate. There exist three sets of assumptions that have attempted to explain and understand the need for local government reorganisation in relation to metropolitan areas. These three are the metropolitan model, which underlies much of the orthodox tradition, the public choice theory, which criticises that traditional model, and the local governance approach, which draws upon, and also criticises, the public administration model without being, in itself, a new normative theory. This chapter seeks to outline the debates on reforming local government by focusing on these three perspectives. While the analysis focuses on the need for an area-wide coordinating authority at local government level (without excluding lower tier authorities' participation) for dealing with air quality management issues, this chapter starts by examining the role of local governments within an urban environmental context.

The role of local government

In the quest of how best to govern urban centres, the role of local governments has been increasingly recognised around the world. At international level, for example, such organisations as the World

Commission on Environment and Development (WCED) or the Organisation for Economic Cooperation and Development (OECD) have acknowledged the importance of local governments for being the best placed for both dealing with local needs (WCED 1987: 17) and managing diverse urban environmental problems (OECD 1990: 43). Additionally, at UNCED, the capacity local authorities have for making sustainable development take place was endorsed in Chapter 28 of Agenda 21 (see UN 1993: 233-234). More recently, at Habitat II, the action plan of the Habitat Agenda has also embraced the significance of local governments by outlining in section D of Chapter IV the need for action on the issues of decentralisation and the strengthening of local authorities and their associations/networks (Hundsalz 1996: 8).

At the same time, the role itself of local governments has developed and changed. This changing role has taken place due to a range of global, national and local forces (political, economic, social) which have challenged the traditional participation and organisation of local governments. At global level, for example, some of these forces include the internationalisation of the economy (i.e. globalisation), the integration of trading blocs, and persistent international debt with its consequential imposed austerity measures for developing countries. At national level, increased efforts of decentralisation and democratisation have also contributed to this change. Finally, at local level, urban growth, politics and the distribution of power, the growing participation of other actors - e.g. civil society, NGOs, the private sector - and local government incapacities and service failures have certainly impacted on the role and functioning of local authorities (see Ben-Elia 1996b: 1-7; Humes IV 1991: 1-12; McCarney 1996b: 6-12 and 1996d: 7-15; Norton 1994: 15). The literature on these emerging forces affecting traditional patterns of local government is substantial in the North but still scarce in the South (see, for example, Ben-Elia 1996a; Bennet 1993a; Borja 1992; Clarke and Stewart 1990; Edralin 1996; McCarney 1996a and 1996c; Morse and Hardoy 1992; Sharpe 1993a; Stewart 1986; Stewart and Stoker 1995a; Stoker 1991).

Converging global and national forces as well as diverse existing types of institutions, behaviour and representation at local level in each country defy broad generalisations on the role and organisation of local government (Marcou 1993: 53; Norton 1994: 15). It is however, possible to depict some recent common directions of change. In the North, for instance, the dominance of national governments during the second part of the twentieth century - i.e. the growth of the welfare state after the Second World War - emphasised the role of local governments as local service agencies. While provoking an increase in local government expenditure, personnel and functions (e.g. health, education, strategic planning), the provision of services seemed the main and only role of local authorities - though in the case of Britain, municipal entrepreneurialisation took place in the nineteenth century (Clarke and Stewart 1990: 3; Humes IV 1991: 2; Sharpe 1993b: 10-11; Stewart 1986: 9). The emergence of the post-welfare model during the 1970s and 1980s based on market theories however, started to conceive local authorities - particularly in Europe - as enablers, partners and facilitators instead of providers of service delivery (Bennet 1993b: 15-17; John 1990: 19; Welch 1997: 19).

Although more recent, more rapid and more traumatic than in the North (Humes IV 1991: 1), there is also a process of change taking place in the traditional role and organisation of local government in countries of the South (Ben-Elia 1996b: 1). Like in the North, much of this recent process of change has come from global forces, state reforms and particular local development problems (Edralin 1996: 4; McCarney 1996d: 5-15). Historically, local governments have usually been neglected within the development effort and have been the 'weak' partners in the governing process in the highly centralised, single-party, or military dictatorship systems of the South. In Latin America, for instance, while the latter has been a dominant characteristic in the majority of the countries, the modernisation of the nation-state during the past two decades has slowly started to give political importance to the role of local and municipal governments. This modernisation process has included such issues as democracy and the decentralisation of state structures and functions (see Borja 1992: 130-133; McCarney 1996b: 6-14; Neira Alva 1995: 32; Satterthwaite 1997: 1682). This process of change in the role and nature of local government in the South is not only about recognising the participation of local authorities - something which was commonly denied in the past. It is also about their key role in implementing policies, directly (or indirectly) providing services, facilitating community participation and, more recently, enabling service delivery and coordinating different political institutions and emerging social groups (Borja 1992: 137-141; Edralin 1996: 17-18; Marcondes 1996: 225-226; Rodríguez and Winchester 1996: 33-34; Stren 1996: 33-39).

The existing literature on urban sustainability recognises local authorities, first and foremost, as essential actors for leading urban centres to some form of sustainability (see, for example, Blowers 1993a; Breheny 1992a; Elkin et al 1991; Gilbert et al 1996; Gordon 1993; Hardoy et al 1992; Haughton and Hunter 1994; Richardson 1992; Serageldin and Cohen 1995; Stoker and Young 1993; Stren 1992b; UNCHS 1996a). There exist three common issues that have emerged from the urban sustainability discourse which are linked to the debate on the changing role and reorganisation of local authorities. First, local authorities are regarded as key agencies for implementing, promoting and designing urban environmental policies. Based on the assumption that it is at local level where knowledge and expertise on local needs and environmental conditions exist, local authorities are viewed as the most appropriate for that task as opposed to central government (Gordon 1993: 13; Haughton and Hunter 1994: 300; Keating 1993: 47). This assumption, however, does not seem to apply for many countries in the South where strong centralised forms of government have limited the role of local authorities to only ensuring certain basic services. For example, in Latin America, because local authorities have barely acted as agents for development, they lack the required technical and political qualifications as well as the resources and the experience in directly managing public services (Borja 1992: 133-134). Although it has been argued that municipal staff do not always have the required expertise to address the conceptual, as well as the practical, issues related to urban environmental policy, training programmes and professional education have been encouraged (see OECD 1996: 167).

Second, while local authorities are not the only key players for managing urban environmental issues, they are seen as one of the many components of the broader concepts of local governance (see Gilbert et al 1996: 16-17; Hardoy et al 1992: 23). The emphasis now given to the positive role of local governments has also included that of community based and non-governmental organisations (Stren 1992b: 312). The importance of transferring powers and resources (e.g. financial, technical) to the local level and the need for adopting an integrated (management) approach to urban environmental problems have also been stressed. Hence, the capacity of local authorities is to work in partnership with other agencies in the pursuit of sustainable development, for example, with the private sector, community organisations, central government bodies, and citizens in general (Hardoy et al 1992: 196-197; Serageldin et al 1995: 1-2). The active participation of communities, for instance, has become essential for achieving urban sustainability and solving urban environmental problems. Being closest to the communities, local governments are regarded precisely as the most appropriate institutional mechanisms for enhancing individual participation within an entire urban centre (see Gilbert et al 1996: 30; Serageldin and Cohen 1995: 29-30). As endorsed in Agenda 21, local authorities play a vital role in educating and mobilizing individuals for increasing people's awareness of sustainable development issues. While developing and adopting 'a local Agenda 21' through a process of public consultation, local authorities gather information from the citizenry to build consensus on urban environmental policies and reshape their sustainability strategies (see Keating 1993: 47; UN 1993: 233-234).

Finally, the changing nature of local government that has resulted in an increased call for local authorities to think of themselves as 'enablers' rather than as direct providers of services has been adopted by the sustainable urban development discourse. Being incorporated into the field of urban sustainability, the recommendations on the issue of the role and organisation of local government are, similarly, for local authorities to increasingly act as coordinators, facilitators and enablers of environmental strategies and policies (see, for example, Hardoy et al 1992: 197; Haughton and Hunter 1994: 300).

Certainly, diverse competing theories and approaches on the role of local government exist which, as argued by Wolman (1996: 158), 'vary from country to country and are embedded in each country's history and political culture'. Unsurprisingly, the number of theoretical traditions and approaches - particularly derived from the North - is quite substantial, and a comprehensive review of them is not possible. As it has been said that countries like Britain and the United States have had the longest and best established traditions of local self-government and representative democracy (Magnusson 1986: 1), much of the discussion on local government has focused on these two English-speaking countries.

The local government reorganisation debate

This section attempts to explore three diverse sets of assumptions of local government reorganisation in relation to the structures of metropolitan areas: the still influential traditional orthodoxy, the public choice theory, and the emerging system of local governance.

The traditional orthodoxy: the metropolitan model

As cities in the North started to expand, concerns were expressed in the latter half of the nineteenth century relating to the political fragmentation of local governments in metropolitan areas. A good example which early in this century provided a diagnosis of the problems arising from metropolitan political fragmentation in the United States is the book 'Metropolitan Government' by Victor Jones (1942). In this work, the author identifies three main aspects that have resulted from a system of disintegrated local government in metropolitan areas: unequalised services, disparity between need and fiscal ability to meet that need (i.e. uneven distribution of tax resources), and dispersion and dissipation of political control of the development of social, economic and political institutions. While the debate on metropolitan fragmentation has largely concerned these three issues, Jones analyses diverse cases that exemplify the scale of fragmentation that existed by the 1930s in the USA. Two of the seventeen metropolitan districts are, for instance, the New York-North-Eastern New Jersey area and the Cook County part of metropolitan Chicago. In the former, within a land area of aproximately 350 square miles, there were around 289 incorporated municipalities, 14 counties, plus the five counties within New York City and innumerable school districts, special authorities and other governmental units. In the latter, the responsibility of local government was divided among 358 separate units: 89 cities and villages, 1 county, 30 townships, 195 school districts, 45 park districts, 1 forest preserve district, 4 sanitary districts, 2 mosquito-abatement districts, and 1 health district. The land area of Chicago comprised around 210 square miles.[1]

By giving a few selected examples relating to the fragmentation of local government, Jones (1942: 52-84) illustrates the effect upon urban life in the USA of what he calls 'the present chaos of governmental units'. One of the many cases that he cites regarding unequalised services and uneven distribution of tax resources, is transportation and traffic management - particularly, the highway and traffic system. According to the author, the problem was that in some of the seventeen metropolitan districts (each containing a population of 750 000 or more by 1940) there were two or more groups of governmental units involved in this issue, under the authority of two or more bodies of statutory law. The ease and speed with which people could move about a metropolitan area depended on highways not only built and maintained by separate jurisdictions, but also on the enforcement of traffic policies by independent police agencies. In the case of the New York-Northeastern New Jersey area, for example, Jones (1942: 57) states that concerted action 'is possible only when two hundred and eighty-four diverse and conflicting units of government are willing to cooperate'. According to the author though, local governments have failed to cooperate in providing traffic facilities mainly because the suburbs know that the central city would be forced, without waiting for cooperation, to provide some degree of accommodation for the daily swing of population in and out of the city. Thus, the suburban motorist would escape the taxes that the central city would have to levy to finance such projects. The costs of the construction of a traffic circuit in a city jurisdiction that would benefit people living outside that area would be borne by the city alone. This fragmented system of authorities led Jones (1942: 57) to assert that such 'a congeries of authorities' was inadequate to meet the problem of providing an articulated system of highways for metropolitan traffic.

During the long phase of local government reorganisation of the 1960s and 1970s in many Western countries, the metropolitan problem was defined as how to achieve the greatest efficiency in production and the greatest equity in distribution. Thus, the urban governmental question referred to 'which organisational and/or procedural arrangement will produce most efficiently and distribute more equitably the services to be rendered locally' (Dente 1990: 59). Solutions to political fragmentation, though, encompassed two diverse methods: those requiring fundamental structural changes in government, and those involving few or no structural changes in existing units of local government. Some examples of the former include the annexing of contiguous territory or the consolidation of municipalities, the merger of special authorities with either the central city or county, and the establishment of a two-tier municipal government for the metropolitan area. For the latter, some examples are the establishment of ad hoc authorities, intergovernmental arrangements, and the extension of state or federal/central administration (see Jones 1942: 85-185).

According to Bollens and Schmandt (1965: 371-399), two different approaches through cooperation among local governments - i.e. with no structural changes in government - have been devised in metropolitan areas in order to confront political fragmentation: interlocal cooperation and metropolitan councils. Taking many forms, interlocal cooperation goes from informal, verbal understandings where administrators of two local governments exchange information on the same service, to the formal, written agreements among diverse local units that jointly decide to build and operate a major service. Formal agreements can be of three kinds: a single government performing a service or providing a facility for one or more other local unit, two or more local governments jointly performing a function or operating a facility on a joint basis, and two or more local governments assisting or supplying mutual aid to another in emergency situations. A metropolitan council is a voluntary, but permanent, association of governments that is convened regularly to discuss and try to agree on solutions to common difficulties and needs. As an area-wide mechanism, it constitutes a forum for deliberation and discussion and an advisory, and coordinating organisation. Neither the interlocal cooperation nor the metropolitan council approaches are imposed by the central government but arise from local institutions.

Certainly, the method of integration by major structural change in government involves more legal and political barriers than the approaches outlined above (Jones 1942: 122). The increased recognition that individual municipal governments were unable to cope with the new social and economic needs of the metropolitan condition resulted in the idea of consolidating all local governmental units into a single jurisdiction for each metropolitan area (see Barlow 1991: 28-36). Although consolidations of local government existed in the late nineteenth century (e.g. the London County Council (LCC) and New York City), it was not until the 1960s and 1970s that the first great movement towards consolidation took place in Europe and North America (Keating 1995: 118).

Much of the discussion and principles for consolidation had their origins in the United States with what has been called the 'good government reform movement' - also known as the 'political reform tradition' or simply the 'old reform tradition' (see Bish 1971: 148; Bish and Ostrom 1973: 7). This reform movement - which dominated the thinking and recommendations of most urban government analysts in the USA until the 1960s - was about the modernisation of local government structures in order to tackle the economic and social challenge of the twentieth century city. Some examples that have explained the principal recommendations of this reform tradition include Anderson (1925 and 1934), Anderson and Weidner (1950), the Committee for Economic Development's 'Modernising Local Government' (1966), and two of the pioneers on formulating the concept of metropolitan government, Jones (1942) in the USA, and Robson (1939) in Britain.[2]

In the first quarter of this century, Anderson (1925: 641-642) identified the organic principles of the traditional reform movement: the complete consolidation of the local government, the short ballot, the unification of powers, the separation of functions and the centralisation of administrative supervision (either by a city manager, a commission or an elected mayor). In briefly explaining each of them, the author stated that despite the many disagreements regarding details and the varying personal preferences among forms of government, the reformers were in considerable agreement upon those fundamental principles.

The first of these principles - as explained by Anderson (1925) - states that in each unified urban area there should be only one unit of local government; in metropolitan areas, though, some powers of local self-government may well be left to the other local units. The latter eliminates much overlapping and duplication of effort. The second principle indicates that a consolidated unit of government simplifies the problem of the voter, i.e. centering the responsibility on a single governing body rather than on many local units. Here, it is assumed that governmental organisation becomes simpler, more visible, and more responsible. Anderson continues to explain that there is no perfect unanimity among all reformers upon the points relating to the third principle, unification of powers, which is linked to the two other remaining principles, i.e. separation of functions and centralisation of administrative supervision. On the one hand, while most reformers advocate the abolition of the separation of powers in local government, on the other hand, they recognise the need to separate the functions of legislation and control from that of administration: politics must be separated as much as possible from administration. In other words, while the political or council branch should exercise control, the administrative one should be allowed to do its work without political interference. In order to achieve this, three different proposals or plans for controlling the city government exist. Two of them - the commission and the city manager - unify the powers of government in municipal affairs by means of restoring a council with complete control over the government. The other - the strong-mayor plan - seems to be less favoured as it implies leaving the entire power of local government in the hands of an elected mayor. Nevertheless, whereas most traditional reformers advocate an administration organised under the hierarchical principle, they tend to abandon the commission plan and favour the manager plan. This is because the former not only fails to separate politics from administration, but does not allow a completely unified administration: it has five heads instead of one. By contrast, the manager plan enables the city council to select the most capable administrative head to be held singly responsible for the entire administration.

During the post-war period, the orthodox reform movement, which advocated the case for consolidation, increasingly resembled what in the USA has been called the 'metro scheme' - also known elsewhere as the 'metro model'. Whereas consolidation usually implied eliminating and replacing existing governments by a single city-wide government unit, 'the metropolitan model' or 'two-tier approach' started to be associated with the idea of an area-wide metropolitan authority sharing powers with smaller local governments within its area (Self 1982: 61). This idea of a metropolitan government with an upper and lower tier achieved widespread consensus during the post-war period (see Hicks 1974; Miles 1970; Norton 1983; Robson and Regan 1972; Sharpe 1995a). In this sense, metropolitan government or the two-tier system of government means the creation of a new governmental unit with jurisdiction over the whole metropolitan area while retaining smaller governmental units for local areas. Each level of government has its own functions and responsibilities: whereas the area-wide unit has control over those activities which are best suited to large scale management, the local units deal with those which demand small scale management close to individuals. This system captures the advantages of consolidated government without incurring its flaws: public services for which there are scale economies can be produced efficiently, spillovers are eliminated, services which demand area-wide coordination can be organised effectively, and the city's tax base can be used equitably (Barlow 1981: 122-126).

Both, theoretically and in practice, the metropolitan model has numerous variants or types of solutions for political fragmentation (see, for example, Anderson and Weidner 1950: 192-200; Barlow 1981: 119-178; Dente 1990: 59-71; Jones 1942: 122-154; Jones and O'Donnell 1980: 546-549; Leemans 1970: 158-165; Norton 1983: 3-51; Self 1982: 66-78; Sharpe 1995b: 17-20). So, for instance, in the United States, Bollens and Schmandt (1965: 440-448) identify three variations of the two-tier formula solution: the metropolitan district, the comprehensive urban county plan, and the federation. The first represents a governmental unit encompassing a substantial part or all of the metropolis but generally authorised to perform only one function or a few closely related activities of metropolitan character. As the latter represents a limited functional (but not territorial) role for a metropolitan authority, it is the mildest variation with respect to the concept of metropolitan government. The second calls for the simultaneous transfer of selected functions from local authorities and sometimes other local units to the county government. The functional shifts are comprehensive in scope and occur at the same time: the county assumes functions of an area-wide nature and the rest of the local units (municipalities and other local governments) remain in existence to perform local services. Finally, the third implies the creation of a new area-wide government. Here, the local authorities or municipalities continue to exist, perform local functions for which the metropolitan government is not responsible, and retain their governing boards. Additionally, provisions for local representation in the lower tier exist.

Some disagreement exists on whether a formula in which a metropolitan-wide authority that has been created simply by extending the boundary of the existing core city authority can be included under the metro rubric. For instance, Sharpe (1995b: 12-20) explains that such a formula - also known as the 'unitary version' - can be an example of the metropolitan model. The author argues that, in fact, it may more closely resemble a more normal metro model as it may have a second tier in the form of neighbourhood councils, such as Oslo and Winnipeg. He recognises, though, that the unitary type is not an option that is easily open to the very largest cities simply because of scale. Thus, for this author, the degree of restructuring the government in practice varies from an area-wide body based on voluntary cooperation between existing units of local government in the metropolitan area, to the entirely new structure with full independent powers. By contrast, Self (1982: 61) asserts that metro theories reject the concept of simply expanding the boundaries of a major city (big urban areas with populations of perhaps a million or more) as this is assumed to be 'politically impracticable and democratically undesirable'. Although he recognises that the structure of the metro authority can vary, if this is no more than a coordinating committee of the other local governments with no or very few executive powers (i.e. a system of intermunicipal cooperation), then it is only a step toward metro government rather than its achievement.

Additionally, determining the physical limits of a metropolis is a very contested issue. Territorial extension or number of inhabitants represent arbitrary indicators for considering any city a large metropolis. Gottman (1995: 1) illustrates this by saying that in 1900 an agglomeration of one million people or more was considered a large metropolis: there existed one or two dozen of these. In the 1990s, there are about 200 such agglomerations and in the really large metropolis agglomerates ten million inhabitants or more. According to Bollens and Schmandt (1965: 6-7) a metropolitan area consists of heavily populated land whose central city (or cities) and suburban or outlying parts have a high degree of economic and social interaction. These authors explain that although most people and organisations that have studied metropolitan areas agree that those are the basic characteristics, there is disagreement over what specific criteria should be applied to determine the boundaries of the metropolis. Overall, they conclude that a metropolitan area is a unit in an economic and social sense, but not in a governmental one.

While there might be some difference of opinion on which formula (functional and/or territorial) can be encompassed under the metropolitan model, the two-tier approach has been the most popular, appealing to traditional reformers in order to overcome the problem of political fragmentation (see Barlow 1981: 128 and 1991: 23; Bollens and Schmandt 1965: 439-440; CED 1970: 44-46; Gunlicks 1981: 14; Hallman 1977: 268; Sharpe 1995b: 18). For some time, metro schemes have represented the vanguard of more general movements of local government reform in countries such as Britain, Canada, France and the US (Self 1982: 78-79). In the United States, for instance, although practical applications of metropolitan governments have been few in number (Barlow 1981: 148), the two-level concept received increased advocacy and use during the 1960s (Bollens and Schmandt 1965: 439) where the need for both a community and a metropolitan level of government was recognised:

> In principle a governmental system for America's metropolitan areas must recognise the need for both a community level and a metropolitan level of government...To gain the advantages of both centralisation and decentralisation, we [Committee for Economic Development] recommend as an ultimate solution a governmental system of two levels (CED 1970: 19).

In Britain, the Herbert Report (RCLG 1960) and the Redcliffe-Maud Report (RCLG 1969) suggested the reorganisation of local government in London and in England, respectively, by advocating the implementation of the two-tier level approach in some identified metropolitan areas. Thus, for instance, the Redcliffe-Maud Report concluded that:

> In the special circumstances of three metropolitan areas around Birmingham, Liverpool and Manchester, responsibility for services should be divided in

> each case between a metropolitan authority whose key functions would be planning, transportation and major development, and a number of metropolitan district authorities whose key functions would be education, the personal social services, health and housing (RCLG 1969: 1).

Peter Self's (1982) important work 'Planning the Urban Region', has explained the rationale and principles of striving for a two-tier system of metropolitan government. For this author, the metro model implies the creation of a metro authority which shares powers with smaller local governments within that area where, if necessary, the existing local governments can be left undisturbed or only reorganised at a later stage, as in the case of Metro Toronto. The metro government must have substantial direct powers and must be elected indirectly from leaders of the smaller local units or directly by citizens at large, or through some mixture of the two. Direct election gives more authority and greater independence over policy formulation, but may be negated by stronger resistance to the policies.

Self (1982: 62-66) identifies five issues about the desirability of metro systems: competition and equality, community and lifestyles, area and functions, efficiency and planning, and democracy and accountability. First, he argues that local government units are shaped by political history and thus cannot grow or shrink like business firms by competitive action. This statement comes as a response to a criticism that some writers (e.g. public choice theorists) make in order to justify a fragmented local government system on the grounds that the latter gives citizens a choice between diverse services and taxes offered by each competing unit, and then a citizen can choose his/her place of residence partly because of this. For Self, freedom of choice over local government services (produced by competitive units) may still occur among the various local units under a metro scheme. The delivery of public services are anyway mainly correlated with the wealth of an area and are probably only a minor factor in the individual's choice of residence. At the same time, a metro government seeks to promote equity, by taking over functions, such as transportation, which have broad catchment areas, and others, such as social welfare, which have a very uneven incidence of need. Under this scheme, financial equalisation can operate for the benefit of its poorer and needier units. A metro body has greater local knowledge and accountability than other potential bodies for such equalisation, such as central or state governments.

Second, due to the significance that local communities and their lifestyles (within localised units) have acquired for local government, Self argues that the best way to recognise both the local and metropolitan importance of a service is to divide it between two levels of government in a logical manner. Some examples of such divisions are highways and local streets, main drainage and local sewers, refuse disposal and refuse collection, and so on. In relation to these examples, the author says that:

> Metro systems can recognise the geographical and political logic of these functional divisions, which correspond to differences in scale and externalities between the functioning of a locality and of a large urban system and to the different political interests of the smaller and larger urban community (Self 1982: 63).

One crucial aspect that the author highlights in relation to the concept of metropolitan-local division is that integrated functional management can be accomplished. The concentration of a whole function (or as much of it as possible) in the same hands simplifies coordination and reduces boundary frictions within the service in question. Furthermore, such a division pinpoints organisational and political responsibility for the service. However, the author recognises that this may weaken coordination between services. The issue of functional allocation raises the familiar aspects of the conflict between areal coordination (functional splits) and functional coordination (weakening areal coordination). In spite of this, Self argues that metro schemes give variable answers to the problem either by dividing or sharing some functions between the two levels, or by concentrating a whole function at one or the other level. Hence, a metropolitan government may be 'top-heavy', 'bottom-heavy', or 'more or less balanced'.

In relation to the issues of area and functions, Self argues that the combination of a large metro government with small local units allows a mix of the advantages of both concepts. The problem here is one of size of the smaller units. If the metro scheme as a whole is to be 'bottom-heavy' or 'evenly balanced' the author explains that the lower-tier units must have the minimum size and resources necessary for functional efficiency. Smaller size implies fewer powers: if the units become smaller they may correspond better to community issues but can expect few public powers and may become advisory bodies acting as pressure groups upon higher levels of government. In the case of the London reform of the 1960s, for example, it was sustained that a local borough needed a population of about 250 000 if it was to wield strong powers, especially in the locally important services of education, health and welfare (see also RCLG 1960).

The fourth element relates to efficiency and planning. Here, Self asserts that some of the favourite arguments for reform are the issues of economies of scale and technical efficiency which an integrated local government achieves. While emphasising gains in relation to large-scale infrastructure or bulk facilities or where artificial boundaries inflate costs, the author admits that the gains are often problematic, especially for personnel services where the advantages of employing specialists is offset by the loss of personal contacts and coordination problems. Additionally, he recognises that there are considerable diseconomies of scale due to the high expenditure per head of large units, as experienced by some metro schemes where there is little evidence of overall cost savings. In the case of London, after the 1965 reform, local government expenditure per head grew at a faster rate than for the country as a whole. Nevertheless, Self explains that the improvement in services may have been relatively greater after such reform which encouraged the councillors and officers of new London boroughs to develop and improve services. As the general theme of the book, he also points out the value of overall planning related to the functional problems of the urban region as a strong argument for adopting a metro scheme.

Finally, Self makes reference to the issues of democracy and accountability. He starts by arguing that if local government remains fragmented, then the responsibility for dealing with broader urban problems passes inescapably to central or state governments. Such has been the case of London after the abolition of the Greater London Council (GLC) in 1986 where the fragmentation of London's government has been accompanied by increasing centralisation achieved through institutional reforms, central government appointments, financial controls, and so forth (see, for example, Newman and Thornley 1997: 967). Such a fragmented situation represents for Self a failure of democratic accountability as traditionally understood. In relation to this, the author disagrees with the idea that because urbanisation has become so amorphous, then the construction of viable political institutions cannot be accomplished. He is also opposed to the fact that as urbanisation is so extensive, service provision then becomes the primary responsibility of state governments or regional bodies answerable primarily to central governments. Furthermore, the author objects to the argument that a metro government cannot be democratic if it increases electoral confusion or indifference and opens up a government still more to the influence of special interests. On the contrary, Self asserts that metro systems solve problems of functional division and of political conflict and apathy: they rest upon the geographic logic of democratic responsibility. The democratic argument for a metro authority remains thus, logical and strong on its own terms provided it corresponds to a genuine arena of common problems and interests.

Certainly, some difficulties in the implementation of a two-tiered local government system exist. Some of the more common are the size and boundaries of the metropolitan authority and the lower-tier units, functional allocation (i.e. designating services to the upper and/or the lower tiers), and intergovernmental relationships. These problems have not only been recognised by the advocates of a two-tier system of government but by some of its critics, mainly public choice theorists (see, for example, Bish 1971: 156-157; Bish and Ostrom 1973: 14-15; Bollens and Schmandt 1965: 488; Jones and O'Donnell 1980: 547-548).

In his illustrative work, 'Metropolitan Government', Barlow (1991: 24-27) explains that resolving those two-tier problems involves further application of the principles and criteria for local government reorganisation. In relation to the issue of size and boundaries, the author asserts that, on the one hand, the metropolitan authority needs to be large enough to contain the metropolitan system (so that area-wide functions can be performed effectively), and its boundary needs to take into account patterns of interaction between the metropolitan centre and its surrounding suburbs. On the other hand, the size and boundaries of the lower-tier units must be determined by the functional and community requirements of the most important local services. For example, functional and community principles can be used to establish a minimum and a maximum size, respectively. The units need to be large enough to perform a wide range of functions, but small enough to serve the interests and requirements of localities within the metropolitan area. While it is desirable to establish lower-tier units of similar size to ensure comparable capabilities and resources, their boundaries need to include patterns of interactions associated with issues such as shopping centres, community facilities and so on.

In relation to other problems, Barlow argues that when considering the issue of functional allocation, it is necessary to split such functions between the two levels of government. Here, it is important to consider the possibility of transferring functions from higher-levels of government to the metropolitan authority. Designating services to either the upper or the lower-tier represents one of the most difficult problems to be solved in the two level system of government. The question of size has become an essential element for determining the latter. Some authors who have developed a useful procedure and criteria for assigning functional responsibilities include Bollens and Schmandt (1965: 308-338), Hallman (1977: 177-190), and Hirsch (1964: 332-338). By way of illustration, Werner Z. Hirsch (1964: 332-336) has addressed this question by considering four separate issues as a guideline to help on the distinction of which urban government services can best be performed on an area-wide or local basis. These issues include economic considerations (scale economy), political considerations (people-government proximity), administrative considerations (multi-functional jurisdictions to solve conflicting interests), and welfare considerations (financial arrangements). What can be learned from the analysis proposals of this author, is that all services need to be considered one by one based on certain criteria to determine which functions can be carried out more efficiently on a large scale rather than on a small scale, and vice versa. Moreover, when the picture is not clear, then some functions may operate more efficiently if both tiers of government operate concurrently. However this procedure may help for allocating functions on a small or area-wide scale, the author asserts that:

> ...each situation would have to be analysed on its merits and the advantages and disadvantages of alternative arrangements investigated and compared. In general, financing could be done by the same unit that provides the service (Hirsch 1964: 337).

Regarding the basis of representation for the metropolitan authority, Barlow (1991) explains that an area-wide government can be either comprised by representatives from the lower-tier units or directly elected by citizens. The former is easier to set up, does not over burden the electorate, and reduces ideological differences between the two levels of government. The latter leads the government to make policies in the metropolitan interest without inter-local rivalries and conflict. Inter-governmental relations can follow three different parths. First, lower-tier units may subordinate to the metropolitan authority; second, lower units may be superior, i.e. primary agents of local government; and third, the two levels may be complementary. Depending on the nature of representation at the metropolitan level and on the degree of local autonomy, inter-governmental problems may increase or decrease and thus it can be determined whether or not shared functions can be performed effectively.

In analysing Self's five elements for a two-tier system of government within the context of the problem of air pollution, it is important to determine if they are relevant or not for ensuring an adequate air quality management system. If they are, then the normative prescriptions of the metro model need to be taken into account when reorganising local government structures in metropolitan areas. At this point, it is essential to bear in mind that controlling air pollution suggests taking an urban-wide approach over the whole 'air-shed' or 'air-basin' of a metropolitan area (see Bish and Ostrom 1973: 24; Bollens and Schmandt 1965: 330).

To begin with, the issue of competition in relation to air quality management cannot be taken as a relevant case to argue for either a fragmented government or a metro system. This is because of the nature of the problem of air pollution. Citizens cannot have a choice of residence among diverse local units in a metropolitan area because air pollution may exist over the entire area regardless of existing administrative boundaries. Air pollution in an urban area usually expands beyond the limits of local units and sometimes, beyond the boundaries of the metropolitan area itself. Thus, competition among fragmented or second-tier local units in the case of the metro scheme may prove useless as local units are helpless to protect themselves against the failure of their surrounding neighbours to control it (see Bollens and Schmandt 1965: 329).

More relevant is the issue of equity and financial equalisation in relating to air pollution control. Here a metro system, as opposed to a fragmented one, promotes equity (and can provide financial aid) among those components of an air quality management system that have broad catchment areas, such as emission inventories, monitoring systems and emergency plans. So, for instance, the ideal monitoring network for air quality management consists of an automated network measuring all major pollutants and providing continuous up to the minute data to a central control (see Elsom 1996: 71). Due to the high costs of real-time monitoring equipment, a metro body can operate a system of financial equalisation benefiting the poorer local units. Although financial equalisation may be operated by the central government, the issues of local knowledge and accountability of lower and upper tier authorities may be greater than that of central government. The latter is particularly relevant in those countries where diverse air pollution functions (such as monitoring) have been traditionally allocated to local authorities. There is no reason to believe that in practice a metro authority would show less political impartiality for financial aid than the central government.

The issues of community and lifestyles play not only a major role in managing air quality but also in determining which aspects have a metro or localised character within an air quality management strategy. As metro schemes recognise the importance of area-wide and localised functions (i.e. the concept of metropolitan-local division), it may be the case that lower-tier authorities could share or have their own functions for certain components of an air quality management system. This is certainly true regarding air pollution control policies for better public transport and improved traffic management policies. The analysis of community organisation and urban mobility in different local units within a metro area assists in deciding whether public transport is required or not for that local unit. If it is, then it also helps to decide which tier of government can provide this and what type of public transport this should be. For instance, there are some types of areas which due to their particular characteristics (e.g. residential location, low urban mobility), do not require the expansion of some means of public transport, such as underground or trains. Instead, a system of local buses can operate in these areas and fulfil the public transport needs. While short-distance routes may be provided and/or coordinated by the lower-tier authority (e.g. by a system of local buses), long-distance routes which require an area-wide coordination mechanism (as they may cover two or more local units) can be provided by a metro body. The creation, enlargement, or improvement of existing means of public transport is also very much associated to the socio-economic structure of diverse local areas within the whole city as it relates to the main source of air pollution, i.e. car ownership and car use. The lifestyles of some richer or busy areas may induce a far greater use of private cars than poorer or more isolated areas within the same metropolitan area.

Another example that allows recognition of a metropolitan-local division of a service is the case of air quality monitoring sites. Here, the upper and lower units can concurrently collect data through similar or dissimilar methods of measuring pollutants: e.g. automated or manual systems. It may also be the case that local units can concentrate on certain aspects of the whole function. So, while the area-wide unit may concentrate on measuring pollutants which represent a situation of unacceptable health risks, the local unit may concentrate on those whose levels are acceptable but where there is still a need to continue monitoring them to be assured they will remain so (see Elsom 1996: 68).

Most importantly, and in connection with the issues of area and functions, the arguments for integrated functional management embedded in a metro scheme as explained by Self (1982) match the prime aim of a comprehensive or integrated approach to air quality management. Indeed, the concentration of diverse aspects of this function in a metropolitan unit - e.g. air quality monitoring networks, emission inventories, air quality standards and public information bands, emergency plans and so on - enhances coordination and standardisation. An integrated response to air quality management in the case of air quality monitoring networks, for example, needs some central control. Without a central head, the information provided by local units through diverse monitoring systems will fail to provide an urban-wide air quality situation, which, in turn, is required for the implementation of an emergency plan when rising levels of pollution pose a threat to human health. Such a comprehensive response to air pollution may improve if other determinant functions like public transport and traffic management issues are also coordinated by the central or area-wide unit.

Thus, according to the orthodox model, the issue of air pollution can be a good example of a 'top-heavy' metropolitan authority where the concentration of a whole function is more evident at the upper level than at the lower. It follows that, if this is the case, discussion on the size and financial resources of local units may not be essential. Nevertheless, if local authorities are expected to participate on a concurrent functional basis or by means of implementing and enforcing air quality management measures (such as informing the public, stopping cars, or running their own system of local buses), then size and finance become significant issues. Certainly, whether a metro scheme for air quality management may be 'top-heavy' or 'evenly balanced', depends not only on the political and organisation weight accorded to or accumulated by each governmental level (see Self 1982: 64). It may also depend on whether local authorities (particularly lower tier) are expected to channel citizens' air quality management concerns and demands into urban environmental policies.

The benefits of economies of scale and technical efficiency that can be achieved through a metro scheme as argued by Self are present in the case of air pollution control (see also, Hirsch 1964: 333). Due to the needs of a large-scale air quality management infrastructure, a metro scheme can exploit large scale economies. Thus, far from being unimportant, large scale operation may bring, at least in theory, more efficiency. While measuring efficiency is often difficult (see Travers et al 1991b: 3), it can show, for example, how much automated air quality monitoring sites cost and whether or not there can be overall cost savings for the entire area.

In relation to the issues of public participation and scale economies, Hirsch (1964: 332-338) argues that while air pollution control benefits from major economies of scale - when the populations exceeds 50 000 - 100 000 - it does not require close proximity between people and government (see above). This urban issue resembles a case where, while an effective dialogue between citizens and authority can encourage responsible government action, it can also lead to chaos and often to inaction. Here, the advantages of close proximity do not seem to outweigh the disadvantages. It is important to note though, that Hirsch may not have thought about the importance of citizen participation (people-government proximity) in relation to car ownership and car use as at the time he published his article, motor vehicles were not as yet the main source of pollution. Regardless of this, major benefit spillovers could anyway be expected from air pollution control; these can extend to an entire basin, sometimes covering a metropolitan complex plus the surrounding countryside. The size of the spillover area, in turn, indicates the proper unit for fiscal interrelation; income distribution does not play an important role. Topography and population distribution favour air pollution control to be assigned to a government unit on an area-wide basis. In a similar way to air pollution control, Hirsch considers that transportation enjoys major economies of scale. Nevertheless, proximity has mixed benefits as citizen participation enriches democratic procedure but at the same time prevents decisive socially desirable action from being taken. So, political proximity can and cannot be considered essential: in the absence of clear-cut proximity advantages, scale economies may take more importance. Again, major benefit spillovers could be expected and income redistribution plays only a limited role.

Finally, without a city-wide coordinating authority at the metropolitan level, the responsibility of area-wide functions may pass to central government or any other non-governmental area-wide bodies. While diverse functions may go to central government in a fragmented situation and thus represent a failure of local democratic accountability as explained by Self, such failure may increase if those functions are taken over by ad hoc non-governmental authorities. The question here is whether democratic accountability is relevant for managing air quality. In principle, there seems to be a positive relationship between democratic institutions and environmental protection. This is because with democratic systems there can be a relatively quicker response to air pollution problems than with authoritarian or non-democratic systems. Additionally, democracy can ensure access to air quality information and allow for more open ways of public policy making (see Barry 1996: 116). As already seen, it has however been argued that a close proximity between people and government is not essentially required for controlling air pollution mainly because while citizens' participation may be desirable it can frustrate action (Hirsch 1964: 337). In relation to this, if it is true (as Self argues) that with a genuine arena of common issues and interests problems of political conflict and apathy may be solved, then the democratic argument for improving air quality management strategies through a metro authority becomes relevant.

The public choice theorists

After decades of approaching urban government through a conventional public administration model, a different logical-deductive model emerged during the 1970s, providing guidance for policy analysis and normative recommendations for reform, as well as offering a distinctive way of explaining how governmental agencies behave (Dunleavy: 1991: 147). This new model - which uses concepts and methods derived from economics to understand political occurrences - is known in various ways as collective choice, rational choice theory, social choice theory, mathematical political theory, but most frequently, public choice theory (Dunleavy and O'Leary 1987: 75).

Public choice developed primarily in the United States but its influence has spread worldwide, particularly to other English-speaking countries (Self 1993: 1). The main principles and assumptions of this school of thought are usually associated with the early works of Buchanan and Tullock (1962), Downs (1967), Lindblom (1965), Niskanen (1971 and 1973), Olson (1965) and Tullock (1976). The discussion that follows is based on the writers that have attempted to apply public choice theory to the analysis of local government in metropolitan areas, mainly US authors such as Bish (1971), Bish and Ostrom (1973), Ostrom et al (1961), Ostrom and Ostrom (1971), and Tiebout (1956). It is important to note though, that while public choice theorists discuss orthodox views about local government, such an approach 'tends to generalise excessively from American experience, rooted in a specific set of assumptions and historical traditions' (Keating 1995: 127).

Bish and Ostrom (1973: 18-21) identify three sets of assumptions within the public choice theory: assumptions about individuals, about goods and services, and about organisations. First, individuals are assumed to be self-interested, rational and maximizers, possessing information about their preferences which can be perceived, ranked and compared easily. Second, there exist purely private goods (highly divisible that can be provided under competitive market conditions), purely public goods (highly indivisible where once provided for some they can be enjoyed by others), and an intermediate continuum (goods that involve spill-overs not isolated or contained within market transactions, e.g. air pollution). Third, government organisations are not only the means for individuals to communicate their own preferences through such mechanisms as elections, but for ensuring that individuals contribute their share for the provision of goods and services (e.g. through payment of taxes). Here, the question to solve becomes one of getting the best results by having all public goods and services delivered either by a single integrated bureaucratic structure subject to the control and direction of a single chief executive, or by having access to a number of individual collectivities capable of providing them in response to a diversity of communities of interest (Dunleavy 1991: 3; Ostrom and Ostrom 1971: 203-206; Ostrom et al 1961: 833).

Public choice theorists argue that urban needs cannot be adequately met by a simple consolidated system, nor by a two-tier system of government. Rather, they recommend a governmental system of multiple, overlapping jurisdictions which can take advantage of diverse economies of scale for different public services. Furthermore, they argue that a public economy composed of multiple jurisdictions is likely to be more efficient and responsive than a public economy organised as a single area-wide monopoly (see Bish and Ostrom 1973: 2). According to Dowding (1996: 53), most of the public choice literature recommending small jurisdictions derives from Tiebout's model of small-scale units in metropolitan areas competing for 'consumer-voters' by the package of tax and services those units offer. For Tiebout (1956: 419-421), local governments are firms and citizens, consumers. The latter are assumed to be fully mobile and may move to that community where their preferences can best be satisfied:

there is a large number of these communities to choose from. Each community has an optimal size where services are supplied according to the preferences of that community - public services show no external economies or diseconomies on the supply side.

While public choice theorists advocate a system of fragmented local government units, they also recognise that both small and large jurisdictions are needed for providing some goods and services, such as for the development, maintenance and use of a network of streets, thoroughfares and highways which serve diverse communities of interest. Bish and Ostrom (1973: 21), for example, explain that some goods and services may be most efficiently provided by large organisations where economies can be realised by serving large populations and areas, and others by small jurisdictions where diseconomies are likely to occur when goods and services are organised on a large scale (e.g. education, police). Although it is acknowledged that some services are more efficiently provided by large organisations, public choice theory favours the multiplicity of governmental units in metropolitan areas with the consequent various legal and informal relationships which exist among them (see Hallman 1977: 62).

One of the main public choice criticisms of large-scale systems has focused on the orthodox assumption that coordination can be achieved through a single, hierarchically organised, area-wide authority. In relation to this, Bish (1971: 151-152) argues that such a conventional premise is theoretically and empirically false. In his case-study of Los Angeles County, in the American State of California, the author asserts that the problems of area-wide coordination can be dealt with by small political units through cooperative or bargained agreements where there is no imposition from a single, larger, and 'outside' political unit. The author concludes that:

> The observation that communities will cooperate on an area-wide basis when all parties benefit, either from the action itself or from compensation by gainers to losers from the action, is an indication that the polycentric system is capable of dealing with area-wide problems as well as meeting the demands of relatively small homogeneous groups within it - through cooperative effort for mutual gain rather than through an imposed political solution by some outside unit (Bish 1971: 93).

In the same vein, Bish found that many independent municipalities in the Los Angeles County have considerable power to prevent the imposition of costs by other units of government. So, a city can prevent the construction of a throughway that will predominantly benefit users on either side of the city rather than the citizens in areas through which it passes. The problem where high-income neighbourhoods can prevent the imposition of costs on themselves but where low-income ones seem to be ideal places for highway construction and the like, cannot be solved by a single political unit as it may be through the cooperative effort of diverse units. If non-hierarchical relationships provide coordination, then the question becomes, according to this author, one of efficiency in the delivery of goods and services to meet individual preferences.

Again, as in the case for coordination, it is argued that there is no reason to assume that a hierarchical area-wide unit would always be the most efficient organisational arrangement to meet citizen preferences (see Ostrom and Ostrom 1971: 204). According to Bish (1971: 45) the starting point is the assumption that individuals have different interests and that one of the major functions of political organisation is to assist individuals to articulate those interests. This is framed under two basic economic concepts: demand and supply. On the demand side, Bish and Ostrom (1973: 22-26) argue that as individuals' interests vary within an urban community, the problem of having only a single vote to express preferences on a wide variety of issues (through a large-scale unit), is diminished as governmental units become more numerous and specialised in their range of functions. An optimal situation is one in which each of several units performs multiple services. As Cox and Nartovicz (1980: 198) explain, 'fragmentation is viewed as an efficient institutional mechanism for the expression of individual preferences...the individual shops among local governments for public goods in much the same way as he/[she] shops among firms for private goods'.

According to Bish and Ostrom (1973: 22-24), there are different ways in which citizen preferences or demands for public goods and services can be expressed. Some of these include lobbying, public opinion polls, petitions, demonstrations, court proceedings, pressure groups, political party organisations, and so on. Although accessability to them varies, all provide individuals with convenient ways of expressing their preferences on single issues. Among all these diverse ways, much emphasis has been placed on voting. These authors explain that either directly or indirectly, voting offers opportunities to individuals, and even though it has weaknesses (voters do not usually agree with a candidate on all issues so their votes do not reflect all their preferences) they are diminished under an organisational arrangement in which each of several units performs multiple services.

On the supply side, the critique to a large-scale organisation has focused on its monopolistic position when providing certain goods and services. Such governmental behaviour leads to a lack of motivation to innovate, improve, or reduce costs when delivering goods and services: 'monopoly suppliers do not have to be responsive to the demands of those they cater for' (Cox and Nartovicz 1980: 197). Since the same level of production for all goods and services cannot be expected under a fully integrated unit of government under monopoly conditions, it has been suggested that rivalry and competition can alleviate some of the most adverse consequences of monopoly behaviour. If ample fragmentation of authority and overlapping jurisdictions exist, sufficient competition may be engendered to stimulate a more responsive and efficient public economy in metropolitan areas (see Bish and Ostrom 1973: 29-30). Just as happens in the private market, in the local public economy model, public entrepreneurs and citizens seek the best way of providing services through a mixture of cooperation and competition (see Keating 1995: 126-127). By revealing preferences (allocative efficiency), the market produces productive efficiency through competition (Dowding 1996: 53).

Finally, Bish and Ostrom (1973: 30-31) identify several ways in which competition can constrain the monopolistic behaviour of public officials. First, there is political competition, where elections of public office posts are determinant. When responsiveness and efficiency decreases, citizens can vote for new officials which may improve governmental outputs. Second, there is the 'voting with the feet' proposal (Tiebout's model), where individuals dissatisfied with the public goods and services provision can move to other places (e.g. districts, municipalities) to meet their preferences. Third, there is a system of alternative producers of public goods and services. This implies availability of different options for diverse services without moving or changing location. Fourth, there is contracting out, where producers can be public or private agencies. Producers will have incentives to improve the quality of services, innovate, increase efficiency, and reduce costs. Competition between diverse producers ensures that those goods and services most intensely desired will be the ones produced, and that they will be produced at minimum cost (see also Cox and Nartovicz 1980: 197).

Apart from criticising large-scale systems of government, public choice advocates have also rejected the traditional reform proposal of a two-tier formula. Although in principle a two-tier arrangement may enhance efficiency and responsiveness, it has been argued that such a system may be insufficient to deal with the diverse demand and supply schedules for all public goods and services over large urban regions (see Bish and Ostrom 1973: 33). According to Bish (1971: 156-157) the problem is that with a two level structure, local communities surrender veto rights to area-wide functions rather than bargaining among themselves to create mutually satisfactory solutions so as to impede the imposition of political externalities. The assignment of certain functions to specific units limits the opportunities to seek alternative structures for solutions to unanticipated problems because the flexibility of governmental apparatus is restricted. Obtaining agreement on just what is primarily local and what is primarily area-wide constitutes one of the main problems of the two-tier proposal. Overall, the main problem with the metro model is that it is viewed as a deviation from the ideal single-centred hierarchical type of unit rather than as having any underlying rationale.

Criticisms of public choice arguments are related to technical aspects against fragmentation and to a rejection of how economists have attempted to explain political phenomena.[3] By way of illustration, Gunlicks (1981: 15-16) argues that a tremendous intellectual gap separates consolidationists from public choice advocates: the tendency of these theories to ignore some of the major arguments or concerns of the other theory. For instance, on the one hand, public choice tends to sweep over the issues of coordination and equality of services and to ignore evidence that organised interests tend to be middle and upper class in composition (discrediting the personality politics of local elections). On the other hand, consolidationists do not seem to know how to react to diverse economies of scales, learn that efficiency is not necessarily the result of scale, and that bureaucracies may be self-serving and defeat the goals of reformers. In spite of this, Gunlicks (1981: 16) says that, 'it is the model of the free market economy of the public choice advocates that disturbs critics perhaps more than anything else.'

According to Dowding (1996: 59), the problem with the public choice theory is the problem it has in understanding market failure. The use of economic and economic-like tools 'involve some tools which *are only applicable* to market situations; they use a hammer to drive in a screw.' In addition to this, the fact that public choice has been taken up by key intellectuals and pressure groups to draft arguments, policy proposals and speeches for conservative politicians (i.e Reagan in the USA and Thatcher in the UK) has led to further questioning from across all the political spectrum (see Dunleavy 1991: 4-5). Despite the clear neo-conservative bias of public choice though, the surprising thing is the lack of critical challenge to this model.

In his work 'Government by the Market?', Self (1993: 176-197) reviews the influence of public choice thought upon some events in the USA and in Britain in relation to such issues as bureaucracy, democracy, and centralised or decentralised systems of local government in the UK and the USA, respectively. In his conclusion, this author asserts that such an influence has been ambivalent. On the one hand, if the market system is seen as intrinsically superior, privatisation of government services seems the most effective policy. Government attempts to reduce public expenditure and impose priorities from the centre can lead to restrictions on democratic choice and an increased role for the central government bureaucracy. According to Self, this seems to be the direction in which the British government has been moving. On the other hand, a more positive evaluation of political choice, combined with a wish to restrict and disperse the powers of government, favours a decentralised and pluralist system - some public choice theorists would like to reduce the scope of central bureaucracy. For this author, while central bureaucracy may be unnecessarily large, such a theory is weak on understanding the distinctive contribution to a democratic society provided by a bureaucracy trained in strong and impartial values of public services.

Although the public choice criticism of traditional views may, in fact, ignore some of the major arguments or concerns of the consolidationist or two-tier approaches, some of its assumptions allow a more in-depth analysis of the structural response that is required for managing air quality in metropolitan areas. One first point relates to the public choice assumption that the problems of area-wide coordination can be achieved by small political units through cooperative or bargained agreements. Such assumption has two major implications in relation to air pollution control. First, there exists the orthodox view - put forward by Jones (1942) several decades ago - that in such a fragmented system concerted action is only possible if diverse and conflicting units of government are willing to cooperate. As an air quality management strategy includes several functional areas (for example, public transport, traffic management, environmental health, police, and so on), the amount of participating units is likely to be quite numerous. Thus, the bigger the scale of fragmentation is, the higher the difficulty to achieve concerted action. Second, even if diverse units of government are willing to cooperate and coordination is eventually achieved, as they are likely to have different views and priorities concerning air pollution control (i.e. some may be more affected than others), such concerted action may be obtained only through a long-term bargaining process. If an unexpected air pollution episode occurs (i.e. outbreak of high levels of pollution) during or even after such a bargaining process, there exists the risk that, as local units cooperate in a voluntarily basis, some may react slowly or fail to deal with the situation.

Another problem that public choice theorists have identified with the traditional approach, particularly with the two-tier proposal, is that local communities would inevitably surrender veto rights to area-wide functions. It is then argued that if diverse local units bargain among themselves to create mutually satisfactory solutions, they would impede the involvement of a single, larger and 'outside' political unit. This assumption would certainly be true if the upper tier of government in a metro scheme had exclusive powers on the whole air quality management strategy and if it remained as a non-elected unit. Nevertheless, a conventional assumption of metro schemes is that these are, precisely, democratically elected (both upper and lower tiers) authorities. Furthermore, diverse functional aspects that constitute an air quality management strategy show that there exist area-wide, concurrent, and localised functions. While as a whole function, air pollution may be a good example of a top-heavy metro model, there are sub-functions which may be either concurrently allocated or exclusively assigned to lower tier authorities. Some concurrent sub-functions may include air quality monitoring networks and public transport - specifically bus systems. Exclusive powers to lower tier authorities may include traffic management measures (such as stopping cars, or vehicle emission tests) or health programmes for vulnerable people (such as asthmatics or the elderly). Clearly, obtaining agreement on just what is local and what area-wide - as argued by public choice theorists - remains an unresolved problem within some aspects of an air quality management strategy.

The two basic economic concepts (i.e. demand and supply) used by public choice theorists to explain why a hierarchical area-wide unit is not always the most efficient organisational arrangement to meet citizen preferences do not seem to apply to air pollution control. First, on the demand side, while it is argued that citizen preferences can be more precisely indicated in smaller rather than larger political units, it has been acknowledged that there are some problems which are uniformly experienced by everyone, and thus, both large and small units of government are necessary if those citizens are able to express their demands. Such are the cases of the quality of the atmosphere or the conditions of major transportation networks, which seem to appear only as exceptional cases (see Bish and Ostrom 1973: 24-26). Second, on the supply side, the criticism of a large-scale organisation for its monopolistic position does not contemplate that, invariably, an air quality management strategy includes several sub-functional areas which require local unit involvement. In a two-tier proposal, for instance, both monopolistic and competitive practices can take place. On the one hand, a monopolistic situation can occur within those sub-functional areas (such as trains, underground, air quality bands, emergency plan, police) that are exclusively allocated to the metro body. This kind of monopolistic behaviour though, can be constrained (as public choice theorists argue) if political competition exists. According to the traditional orthodoxy, metro schemes include such political competition either directly or indirectly, i.e. elections of public office posts at the upper level. On the other hand, competition among lower-tier units can take place within those sub-functions (such as local buses, stopping cars, vehicle emission tests, parking spaces) that are distributed among each of them. It is important to consider though, that neither monopolistic nor competitive and rivalry practices necessarily ensure improved air quality, policy innovation, increased efficiency, and reduced costs.

The local governance approach

While public choice has definitely become more relevant to the US administration than to the more collective European style of administration, the new ideas of governance have started to shift attention away from a hierarchical or command mode of government action to a cooperative or partnership mode, and from centralised control to decentralised initiatives (Self 1997: 17). This new approach to the study of local government is now being used by international organisations (UNCHS 1996a: 161), and among local government comparative studies (Barlow 1991: 35-36; Gilbert et al 1996: 16-17; Humes IV 1991: x-xii;

McCarney 1996b: 4-6). Furthermore, it has also been used as an innovative way for addressing and explaining current systems of local government in countries such as in Britain (Cochrane 1993: 69-80; Goodwin and Painter 1996: 635-637; Johnston and Pattie 1996: 672; Stoker and Mossberger 1995: 211-214; Tickell and Peck 1996: 595-596).

Current analysis and discussion on the role and organisation of local government thus, is increasingly being included under the all-encompassing term 'governance'. Three diverse connotations of this term at city level exist: 'metropolitan', 'urban' and 'local' governance. One of the earliest references to this concept appeared during the beginning of the 1980s when attempts were made to define the essence of governance in metropolitan areas: 'the structure of relationships among governmental and other organised actors with interests in completing or preventing activities with interjurisdictional impacts' (Jones and O'Donnell 1980: 541). Analogous definitions of this concept have developed over the years under the labels of 'urban' and 'local' governance, all sharing the same underlying idea. By way of illustration, the UNCHS (1996a: 161) notes that 'local governance' is a more inclusive term than local government as it encompasses a wider range of other actors (public, private, individual) as well as their relationships. In the same vain, and within the context of developing countries, McCarney (1996b: 4) explains that the term governance broadens the space to include the essential role played by organisations in civil society where formal structures of the state are weak and unable to provide services. For this author 'urban governance' refers to 'the relationships between civil society and the state, between rulers and the ruled, the government and the governed' (quoted from McCarney et al 1995).

Local governance includes key players (such as ad hoc authorities, NGOs, the private sector, and so on) who in the past were not thought of or simply were not actually involved in the process of governing metropolitan areas. While governance implies less hierarchical and bureaucratic structures of decision-making and more methods based on local networking and negotiation, these new actors have usually been business-oriented:

> Although the traditional conduits for local politics and policy implementation - the local authorities - remain important, increasingly they are having to coexist, collaborate and compete with a plethora of new agencies, networks and organisations, all jostling for local resources, power

> and influence. One of the defining characteristics of these new structures of local governance is that in different ways and to different degrees they are business-led (Tickell and Peck 1996: 595).

According to Self (1997: 17-18), there exist two diverse approaches to governance. The first approach is exemplified by Jan Kooiman's (ed., 1993) 'Modern Governance: New Government-Society Interactions'. Here, government must be updated to match the variety, complexity and dynamism of modern societies where neither traditional bureaucracy nor market-based reform packages are up to the task. Rather, what is needed is the maximum devolution of powers to largely self-regulating institutions or public-private partnerships coupled with the strategic design and motivation of complex inter-organisational networks. Coordinated planning at the centre will help to design and steer the operation of largely autonomous decentralised institutions. The second approach (as explained by Self) is exemplified by Osborne and Gaebler's (1992) 'Reinventing Government'. This work envisages decentralised public agencies co-operating with business, community and voluntary groups to improve public service delivery and meet new or urgent social demands. The emphasis is upon entrepreneurial officials in partnership with other local actors and stakeholders developing a culture of local initiative and problem-solving. As Stoker (1996b: 3) notes, their work 'is about how a government might make sensible and effective use of a wider range of tools beyond the direct provision of services. Governance for them is about the potential for contracting, franchising and new forms of regulation.'

It is this last idea of governance as presented by Osborne and Gaebler (1992) that has been associated with what it is known as the new public management (see, for example, Clarke 1997: 39; Hood 1991: 5-8; Stoker 1996b: 3). The new management of local government discusses the way local authorities organise themselves to carry out its work, how they determine and implement their policies, and how they plan, choose, influence and act. The characteristics of the new management involve a commitment to openness, learning and innovation (Stoker 1991: 236). Unlike the orthodox position which sees local authorities mainly as providers of services, the new management approach considers them both as agencies for the delivery of services and as political institutions with a capacity for local choice. The consideration of structures is a distraction from the real issues of the role and way of working of local authorities. It is only when the nature of local government is known that it is meaningful to discuss such issues as structures, tiers, and boundaries. Furthermore, unlike the public choice theory, it regards local authorities as public sector and not market organisations; that is to say, local authority decisions are subjected to political control rather than market discipline (see Stewart 1983: 1-4; 1986: 2-4 and 1995: 250). The evolutionary nature of public management has recently embraced new emphases and trends (such as focus on performance, disaggregation, empowerment) and new approaches (such as competition and markets) (Clarke 1997: 41-42; Clarke and Stewart 1996: 48). The literature on such an approach has considerably expanded during the 1990s (see, for example, Ben-Elia 1996a; Clarke and Stewart 1990; Davis et al 1997; Leach 1992; Leach et al 1994; Stewart and Stoker 1995a).

While the assumptions of the new local public management and the contribution of the local governance perspective provide a reference point which challenge many of the assumptions of traditional public administration, none of them conform to a coherent or consistent theory (Clarke 1997: 41; Stewart 1986: 3; Stoker 1996b: 4). Rather, the governance approach draws upon diverse theories, such as institutional economics, international relations or public administration, and is regarded as an organising framework for understanding the changing processes of governing (Stoker 1996b: 3-4).

Instead of making a series of statements than can be shown to be either true or false, Stoker (1996b: 1-15) outlines five diverse propositions or aspects of governance for consideration - each with a certain dilemma or critical issue. These include: multi-agency partnerships, a blurring of responsibilities between public and non-public sectors, power dependence between organisations involved in collective action, the emergence of self-governing networks and the development of new governmental tasks and tools. First, governance refers to a complex set of institutions and actors that are drawn from but also beyond government. While highlighting an increased involvement of different agencies in service delivery and strategic decision-making (public, private, voluntary) it challenges conventional assumptions which focus on government as if it were a 'stand alone' institution divorced from wider societal forces. Here, the dilemma of governance is one of legitimacy as there exists a divorce between the complex reality of decision-making associated with governance and the normative codes used to explain and justify government. On pragmatic grounds, to be effective in the long run implies that power holders must be seen to be legitimate. In Britain, for example, the latter has created tension concerning unaccountable quangos, the difficulty of separating policy and operational matters, the influence of faceless bureaucrats and the nature of ministerial accountability. Thus, the issue to be considered is whether, or how, governance can obtain enhanced legitimacy.

Second, governance recognises the blurring of boundaries and responsibilities for tackling social and economic issues. The starting point is that there exists a shift in responsibility, a stepping back of the state, and a concern to push responsibilities on to the private and voluntary sectors, and more broadly, to the citizen. The agencies or sectors that governance recognises include voluntary groups, non-profits, non-governmental organisations, community enterprises, coops. mutuals and community-based organisations, and the private sector. The dilemma suggested by the blurring of responsibilities is that it can lead to blame avoidance or scapegoating. In other words, the latter creates a situation whereby government actors can pass off responsibility to privatised providers or other organisations when things go wrong and then blame others for failures and difficulties. The blurring of responsibilities creates an ambiguity and uncertainty in the minds of policy-makers and citizens about who is responsible for what.

Third, governance identifies the power dependence involved in the relationships between institutions involved in collective action. Power dependence implies that organisations committed to collective action are dependent on other organisations and need to exchange resources and negotiate common purposes in order to achieve their goals. The outcome of exchange is determined by the resources of the participants and the rules of the game and the context of the exchange. In a governance relationship no one organisation can easily command, although one organisation may dominate a particular process of exchange. From the governance perspective, governing is always an interactive process because no single actor (public or private) has the knowledge and resource capacity to tackle problems unilaterally.

Governance as an interactive process involves diverse forms of partnership: principal-agent relations, inter-organisational negotiation and systemic coordination. First, the principal agent form rests on one party (the principal) hiring or contracting another (the agent) to undertake a particular task. Second, the inter-organisational form involves organisations negotiating joint projects in which, by blending their capacities, they are able to better meet their own organisation's objectives. Finally, the systemic co-ordination form sets up a level of mutual understanding and embeddedness so that organisations develop a shared vision and joint-working capacity, which leads to the establishment of a self-governing network. The critical issue in this third proposition is that power dependence exacerbates the problem of unintended consequences for government, i.e. intentions do not always match outcomes. In principal-agent relations the principal does not have complete control over the agent and has only partial information about its behaviour. In the inter-organisational form, negotiated relationships can lead to ambiguous outcomes which can be interpreted appropriately by the various participants. Unintended results, however, are not necessarily undesirable nor perverse.

Fourth, governance is about autonomous self-governing networks of actors. Governance networks imply influencing government policy and taking over the business of government. Under governance, actors and institutions gain a capacity to act by blending their resources, skills and purposes into a long-term coalition - sometimes called 'regime'. A regime can be defined as an informal yet relatively stable group with access to institutional resources that enable it to have a sustained role in making governing decisions. Although participants are likely to have a domain of command power, the regime is formed as an informal basis for coordination and without an all encompassing structure of command. Here, the dilemma with such self-governing networks is that of accountability. This can be seen at two levels. First, members of particular groups may be dissatisfied with the network arrangements agreed by their leaders and yet find it difficult to express or act on the dissatisfaction due to the powerful nature of the network in which they find themselves. Second, even if all members of the groups are satisfied, a problem of accountability can arise because all networks are to a degree exclusive. Networks are driven by the self interest of their members rather than a wider concern with citizen's interests or more particularly those excluded from the network. The solution to this seems to bring government back in some form. While the networks have a significant degree of autonomy (which is required to achieve their purposes), government can indirectly or imperfectly steer them without occupying a sovereign position.

Finally, governance recognises the capacity to get things done which does not rest on the power of government to command or use its authority. It sees government as able to use new tools and techniques to steer and guide. In the context of governance, government has to learn an operating code to challenge past hierarchical modes of thinking. The terms used to describe this new form of governing include, for example, enabler, catalytic agent, commissioner. In this proposition, Stoker (1996b: 14) uses Kooiman and Van Vliet's classification of governance in order to explain the three tasks of government. First, (de)composition and coordination' which involves defining a situation, identifying key stakeholders and then developing effective linkages between the relevant parties. Second, 'collibration and steering' which is concerned with influencing and steering relationships in order to achieve desired outcomes. Third, 'integration and regulation' which involves thinking and acting beyond the individual subsystems, avoiding unwanted side effects and establishing mechanisms for effective coordination (see also Kooiman 1993). The dilemma here is that even if governments operate in a flexible way to steer collective action, governance failure may occur. This is because of existing tensions and difficulties with the institutions of civil society and inadequacies in the organisations that bridge the gaps between public, private and voluntary sectors. Likewise, failures of leadership, differences in time scale and horizons among key participants and the depth of social conflict can also lead to governance failure.

The need for an adequate response by local authorities to urban environmental concerns has been addressed by Stoker and Young's (1993: 5-17) 'Cities in the 1990s'. Their case for local authorities under an emerging system of governance is based on four arguments. First, urban problems require local solutions and local knowledge: local authorities can provide the latter in order to tailor policies to meet particular challenges in their localities. Second, urban problems require an integrated response from a range of agencies and interests. The diversity of players and interests raises the issues of coordination which can be secured through a system of networks. Local authorities are precisely regarded as valuable contributors to such local network-building - as long as they are involved in an active and substantial way. Third, urban environmental policies require strategic leadership and vision: local authorities can provide that leadership balancing the concerns of diverse interests. Fourth, decisions on how to deal with urban problems need to be accountable and legitimate: local authorities can provide a forum for ensuring the legitimacy of urban decision-making.

As an organising framework, the governance perspective assists in identifying diverse governing issues concerning the participation of local authorities in their dealings with air pollution control. As explained in past chapters, in order to secure urban sustainability there is a fundamental need for environmental issues to be tackled in an integrated fashion (which means the inclusion of all sectors of society). While this assumption recognises local authority as one of the many components of a system of local governance when dealing with urban environmental problems, it also implies that an adequate air quality management strategy may benefit from the participation of diverse actors. These participants may include local government units, central authorities, political parties, NGOs, the private sector, citizens, and so on. It follows that due to the increased number of agencies that may be involved in a system of air quality management, the problems of legitimacy, accountability and responsibility (as explained by the governance perspective) are likely to appear.

In the context of local governance, the tasks of government outlined by Kooiman and Van Vliet (quoted in Stoker 1996b) - that is to say, (de)composition and coordination, collibration and steering, and integration and regulation - are essential for managing air quality. This is because the diverse functions and subfunctions in a system of air quality management - particularly public transport, traffic management (motorways), air quality monitoring data, emergency plan (stopping and banning cars) - demand a great deal of area-wide coordination. Hence, in order to build and coordinate local networks, protect and regulate air quality, local authorities are expected to provide strategic leadership. Nevertheless, while political or civic leadership is required in order to balance the concerns of diverse interests, there is a strong tendency for leadership to seek to impose order and issue directives. If local authorities only indirectly and imperfectly steer networks without occupying a relatively sovereign position, or at least some form of command mode of governing, some governance failure may occur. Weak local authorities, for instance, may fail to give leadership and coordination in the implementation of traffic management measures (such as banning cars) when an air pollution episode occurs. Whereas some relative imposed leadership may be required for this and other related issues, other functions and sub-functions within an air quality management strategy may only need steering and guidance.

Conclusion

This chapter has outlined three diverse set of assumptions for reorganising local government: the traditional orthodoxy (hierarchy), the public choice theory (market) and the local governance approach (networks). First, the orthodox public administration model advocates a systematic hierarchy of local administrative bodies, which are relatively autonomous and multi-purpose, and provide a wide range of services. In analysing local authorities in metropolitan areas, it sees political fragmentation as a problem and thus advocates consolidating existing units of local government either by creating a single city-wide unit or by adopting a two-tier system of government called the metro model. With numerous types, the metro model is usually associated with the idea of an area-wide metropolitan authority sharing powers with smaller local units within its area. Second, public choice theorists argue that urban needs cannot be adequately met by a single, hierarchically organised, area-wide authority (consolidated unit), nor by a two-tier system of government. By using concepts and methods derived from economics (i.e. economic-like tools which are applicable to market situations) to understand political occurrences, they recommend a governmental system of multiple jurisdictions which can take advantage of diverse economies of scale for different public services, and be more efficient and responsive than a public economy organised as a single area-wide monopoly. Third, the local governance perspective shifts attention away from a hierarchical or command mode of government action to a cooperative, partnership or network mode. This approach explains that local authorities, while remaining important, have to coexist, collaborate and compete with a wide range of agencies, networks and organisations (e.g. non-governmental organisations). Thus, local authorities need to be reorganised or renovated to match the variety, complexity and dynamism of contemporary societies, which neither the traditional orthodoxy nor market based reform packages are able to do.

In the following three chapters, I analyse the way in which the government of London and Mexico City have been responding to the problem of air pollution. In doing so, I will return to the models of local government outlined in this chapter in the conclusions at the end of this book.

Notes

1 The facts and figures of fragmentation were taken from Jones (1942: 16 and 124). The land area in square miles corresponds to the year 1930.

2 For an extended list of references on the old reform tradition literature, see Bish (1971: 148-149).

3 Some interesting criticisms of the public choice theory include Bollens and Schmandt (1965: 63-67); Cox and Nartovicz (1980: 196-198); Dowding (1996: 59-64); Golembiewski (1977: 1488-1507); Gunlicks 1981: 15-17); Hallman (1977: 62-63); Rose-Ackerman (1983: 55-57); Self (1993: 176-197).

4 Local Government Organisation in London and Mexico City

Yet people wish to know what part Local Government fulfils in the general scheme of our lives, its functions, how it evolved, the principles that govern its conduct, and the problems that face those anxious to make it even more serviceable than it is at present.

Herman Finer (1993) in the Preface of his book 'English Local Government'

Introduction

The geographical area of London has never been easy to determine. London can be as big as the observer or analyst may want it to be: its physical extension could go from Poole Harbour to the Suffolk coast, from the middle of Northamptonshire down to Sussex (Hall 1989: 3-6). Nowadays, London or Greater London refers to a physical area of 1 578 sq. kms., currently covered by 33 local units (32 boroughs and the City of London). With a total population of 6 933 000 the London boroughs (with the exception of the City of London) are evenly matched units with populations in the range of 130 000-300 000 (see Table 4.1). The London area corresponds not only with the statistical and physical (continuously built-up) definition of the metropolis, but with that of its local government (see Map 4.1). Since 1965, the boundaries of London have remained basically the same, and in spite of the 1986 major reorganisation of its governmental structures - i.e. abolition of the GLC - the London government still presents 'a remarkably compact appearance' (Hebbert and Travers 1988: 192; Hebbert 1992: 139; Hoggart and Green 1991: vii).

The geographical area of Mexico City has been commonly identified with the contiguous built-up area regardless of the limits of the Federal District (DF) and/or its surrounding federated state called the State of Mexico (EdoMex). Before 1950, the built-up area of Mexico City was within the territory of the DF, but since then the area has expanded beyond those limits. Thus, Mexico City has been associated with the 'First Quarter' or inner city (four central delegated units within the DF), the

Metropolitan Zone of Mexico City (MZMC) (the DF and 17 conurbated municipalities), and the Metropolitan Area of Mexico City (MAMC) (the DF and 27 conurbated municipalities) (Burgoa 1985: 914; Díaz Díaz and Perló Cohen 1994: 111; Hall 1984: 214-215; Ward 1990: 19).

Table 4.1 Population in London by borough (1993) and in Mexico City by *delegación* (1990)

Borough	Population (000s)	Borough	Population (000s)
Barking and Dagenham	146	Kensington and Chelsea	149
Barnet	305	Kingston upon Thames	138
Bexley	220	Lambeth	259
Brent	248	Lewisham	240
Bromley	292	Merton	175
Camden	181	Newham	226
Croydon	323	Redbridge	233
Ealing	286	Richmond upon Thames	167
Enfield	262	Southwark	229
Greenwich	215	Sutton	173
Hackney	191	Tower Hamlets	170
Hammersmith and Fulham	155	Waltham Forest	220
Haringey	212	Wandsworth	266
Harrow	207	City of Westminster	189
Havering	232	City of London	4
Hillingdon	240		
Hounslow	207		
Islington	175	Total	6 933

Delegación	Population (000s)	*Delegación*	Population (000s)
Alvaro Obregón	642	Magdalena Contreras	195
Azcapotzalco	474	Miguel Hidalgo	406
Benito Juárez	407	Milpa Alta	63
Coyoacán	640	Tláhuac	206
Cuajimalpa	119	Tlalpan	484
Cuauhtémoc	595	Venustiano Carranza	519
Gustavo A. Madero	1 268	Xochimilco	271
Iztacalco	448		
Iztapalapa	1 490	Total	8 235

Source: INEGI (1994b); Newman and Thornely (1997)

Due to a 1993 constitutional reform, Mexico City is now politically and formally regarded as the Federal District (DF) with a physical area of 1 499 sq. kms. (see Constitution 1997). This area (with a total population of 8 235 744) was created in 1898 and is currently covered by a city hall - the GDF - which is territorially and politically divided into 16 delegated units (see Map 4.2). All sixteen local units (with populations ranging from 63 654 to 1 490 499) are hierarchically subordinated to a mayor, head of the City Hall or Federal District Government (GDF) (see Table 4.1). While the MZMC refers to a physical area of 3 399 sq. kms. (with a total population of 14 582 708), the MAMC refers to an area of 4 620 sq. kms. (with a total population of 15 047 685) which corresponds with the statistical definition of the metropolis (see, for details, INEGI 1992 and 1994b).

This chapter describes and analyses two systems of local government (London and Mexico City) and discusses, from a historical perspective, the debates and experiences concerning governmental reform in both urban centres as well as the relationship of their organisational structures with air quality management. The main purpose of this chapter is to pull the two cities into a single scheme of interpretation.

London: the struggle for reform

The geographical area and local government arrangement that exists in current London has been the result of special historical circumstances originating with the transition of British city government from the pre-industrial closed corporations to modern city governments (Sharpe 1995c: 113). Physical limits and entire systems of local government have been since changed or abolished due to socio-economic growth and political interests in order to assure effective politico-electoral control over particular areas (see, for example, Dunleavy 1991: 120-123). In 1991, the Greater London Group at the London School of Economics and Political Science (LSE) succinctly described two hundred years in the history of London government:

> It is easy to assert...that virtually all possible models of structure, have, at one time or another, been proposed. Increased central government responsibility, London-wide authorities with very different boundaries, appointed and elected authorities, urban parishes and large boroughs, partial

reforms and full reforms, and radically different service allocation have all been suggested or tried during the past two hundred years (Travers et al 1991a: 56).

The debate on creating an area-wide government unit for London began in the nineteenth century with the second report by the Royal Commission on Municipal Corporations in 1837. This report emphasised the need for a single and unified system of local government in London. Nevertheless, the Commission did not explain whether such a unit would resemble the then newly-established municipal corporations in the provincial boroughs or whether it would come under the jurisdiction of the central government (Robson 1939: 22-23). While being ignored by the central government, this report found a chaotic and confusing pattern of government with a prevalence of corruption and inefficiency (see Barlow 1991: 49). According to Robson (1939: 54-55) the condition of the metropolis was one of utter chaos and a 'veritable jungle of areas and authorities and a nightmare of inefficiency'. The division of local administration among different districts of London was considered at the time by some commentators as detrimental to the public interest, including such functions as lighting and paving of streets, traffic regulations and public health (Barlow 1991: 49-52; Gibbon and Bell 1939: 3-26; RCLG 1960; Robson 1939: 23-53; Young and Garside 1982: 11-22).

It was not until 1855, though, that the first form of city-wide administration or government in London was established: the Metropolitan Board of Works (MBW). This body (with similar boundaries to those of 'Inner London') has represented not only one of the three metropolitan Londons - the other being the London County Council (LCC) and the Greater London Council (GLC) - but a major step towards metropolitan government (Barlow 1991: 52). Indeed, while the MBW constituted an indirectly elected upper tier with the main purpose of improving Victorian London's sewage system, the elected local units - 38 parishes - had responsibility over local sewers and drains, road maintenance, and pavements and street lighting (Travers et al 1991a: 49). Although allegations of corruption and scandal contributed to the Government's decision to abolish the MBW, Robson (1939: 64-65) has argued that the real cause of its dismissal was that it failed to awaken civic spirit in the minds of London's inhabitants. Being an indirectly elected authority it precluded the possibility of accountability from the central body and of 'watchfulness or interest on the part of the public'.

In 1889, the London County Council replaced the MBW becoming responsible from time to time for services such as the poor law, fire services, housing, bridges and tunnels, building control, health services and education; the new authority had almost the same physical boundaries as those of its predecessor. Later on, in 1899, a system of two-tier government appeared when, within the LCC area, 27 metropolitan boroughs and Westminster City were created therefore taking some of the responsibilities away from the parishes (see Travers et al 1991a: 50). The discussion concerning reforming local government concentrated on the need to create a municipal and representative body for the whole built-up area of London as well as on the issue of functional distribution where certain local government services were not suitable for central authority administration but for smaller authorities. The problem of air pollution was neither explicitly included in the late nineteenth century proposals as a case for reorganising local government nor regarded as an area-wide or metropolitan issue (see Robson 1939: 62-99; RCLG 1960: 12-13). Two years after the LCC emerged, powers to deal with nuisances from smoke were transferred from the police to the health authorities (vestries and district boards) and later on in 1899 to the 28 newly created metropolitan borough councils (Gibbon and Bell 1939: 559).

The long existence of the LCC, which served until 1965, never enjoyed the prestige, status, and degree of citizen allegiance that should have been accrued to the city government.[1] It was always overshadowed by the existence of a more popular City of London which has retained its medieval non-democratic structure and boundaries, thus remaining unreformed and with the same status as a London Borough to this day (Sharpe 1995c: 113-114). The area of jurisdiction of the LCC was not only an artificial creation with no historical basis but it failed to embrace the entire built-up area in London when it was established in 1889. As London continued to expand, soon the LCC's jurisdiction became an ever smaller proportion of the whole city (Regan 1972: 512; Rhodes 1970: 3). In spite of this, the LCC has remained internationally famous among comparative local government studies and has been regarded as 'the grandfather of metropolitan government' (Hebbert and Travers 1988: 174).

During the LCC's existence two official reviews of the structural arrangements for local government in the Greater London area took place, first in 1921-23 and then, in 1957-1960 (Travers et al 1991a: 51). In the first one, a Royal Commission was set up in 1921 to see whether any alterations were needed in the local government of the administrative county of London and its surrounding districts. This Commission (chaired by Lord Ullswater) came to be as a result of the increasing problems for London's government due to the growth of London beyond the boundaries of the LCC. The main concern focused on whether certain services ought to be administered over a much larger area than that of the LCC (Rhodes 1970: 4-5). One of the themes which brought much attention before and during the proceedings of the 1921-1923 Commission were the problems derived from the growth of London regarding two interrelated functions: town planning and traffic and transport (see Barlow 1991: 68-70). The debate on traffic and transport problems, for instance, was surrounded by the need of a Greater London authority in order to increase coordination and efficiency and improve existing conditions (see Rhodes 1970: 7; Young and Garside 1982: 109-110).

Although the majority view of the Ullswater Commission (1921-23) concluded (through one major and two minor reports) that no changes were needed in the areas, status or functions of the authorities in the Greater London area, some of the themes and arguments foreshadowed much later debate about London government (Rhodes 1970: 4-5). During the 1930s and 1940s, though, the debate on government reform diminished and the idea of a Greater London authority was only argued in relation to planning. Nevertheless, one of the most relevant outcomes of the government reform debate during this first part of the twentieth century, was that the argument for the geographical expansion of the metropolitan government brought several new players into the political arena - mainly outer suburbs of London (i.e. counties, county districts, and county boroughs). As explained later on, the issue of widening the spatial context of London significantly contributed to the 1965 government reform (Barlow 1991: 67).

It is important to note, however, that during the 1950s the idea of a metropolitan government for the Greater London area had become obsolete in the context of political agendas. As explained by Barlow (1991: 73), on the one hand, the Conservative party was publicly against any large-scale local government and was pressing for abolition of the LCC and for a reconstitution of larger and stronger boroughs in the whole area. On the other hand, the Labour party realised that any larger metropolitan government than the LCC would jeopardise its electoral control of London. Additionally, the surrounding local authorities had demonstrated enough opposition to the creation of a Greater London government.

Map 4.1 The Greater London Area

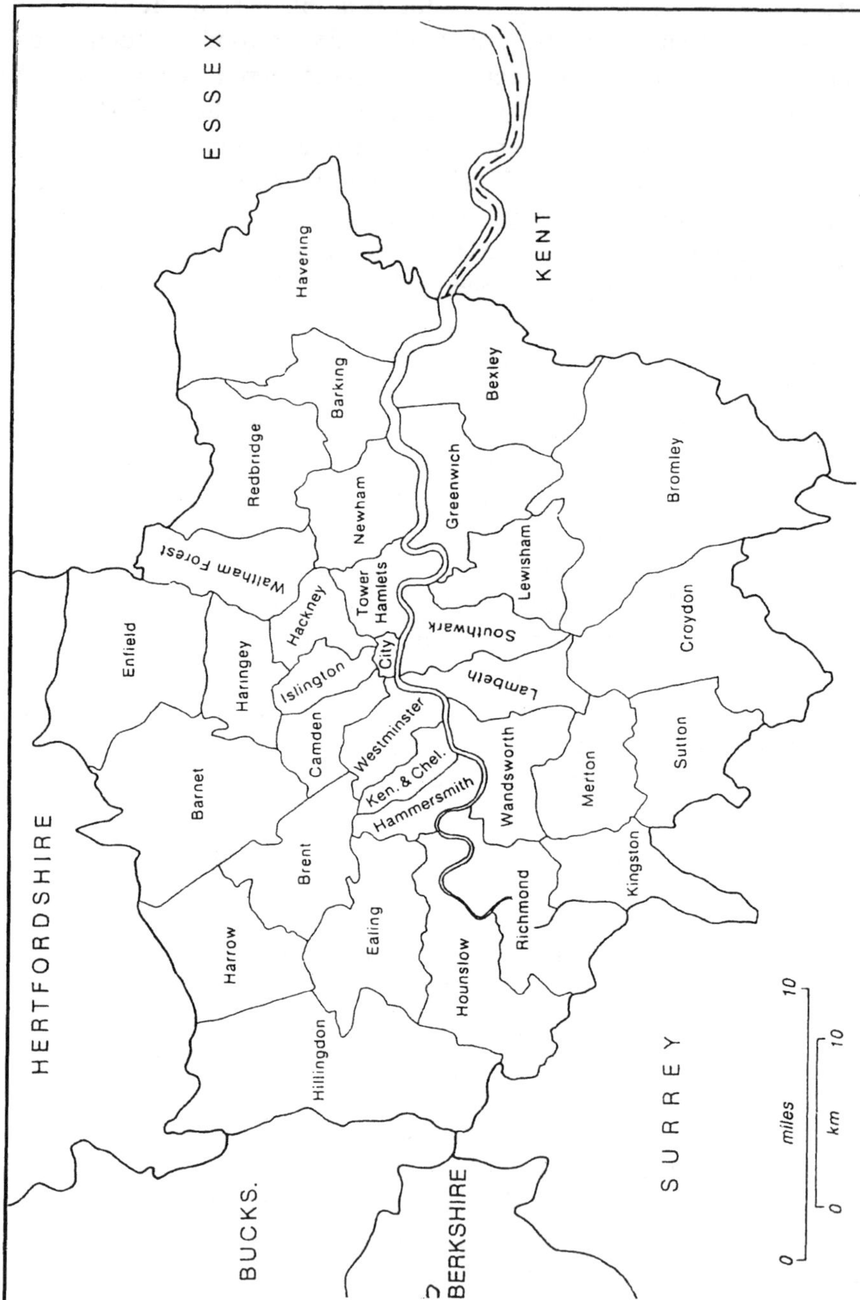

Thus, with such a political scenario, the Government's announcement in 1957 to appoint a Royal Commission to look into the problems of Greater London came as a surprise. While the government unexpectedly showed concern for the effects of London's growth on local government areas outside the county in 1957, particularly in relation to Middlesex (see Rhodes 1970: 17), it was largely 'the Conservatives' political ambition in the 1950s to strike a blow at Labour's LCC heartland which led to its demise in the [1963] London Government Act' (Hebbert and Travers 1988: 174). Hence, it was both the county borough problem and a secret plan by the Conservatives to improve the party's electoral prospects in the metropolis, which led to the establishment of the Royal Commission on Local Government for Greater London (RCLG), also known as the Herbert Commission (see Barlow 1991: 73-76; Young and Garside 1982: 298-317).

The 1960 Herbert Report proposed the creation of a Greater London Council as an upper tier covering an area six times larger than the LCC, and of 52 London Borough Councils as a second or lower tier of local government within the same GLC area. While the upper tier would be responsible for education, planning, main roads, refuse disposal, the fire service, ambulance service, traffic management and research, the lower tier would be for personal social services, environmental health, local roads and libraries. In addition, both tiers would have concurrent powers for housing, recreation, arts, sewage and land drainage. This time, the issue of air pollution was included in the Herbert Report for reorganising local government, although it remained a localised aspect (see also, Rhodes 1970):

> The efforts of one authority to control the air in its area may be to some extent nullified by lack of smoke control in an adjoining area; and it is for consideration whether there is a need for a more comprehensive approach to this problem and whether it should be dealt with on a wider basis. On the other hand the implementation of a clean air policy is essentially local (RCLG 1960: 174).

The recommendations for government reform by the Herbert Commission had two underlying principles. First, the need for a local authority to attend the common problems and needs of the entire metropolis (i.e. administrative efficiency). Second, the desirability of revitalising local government at a lower level (i.e. the health of local government) by increasing the responsibilities of the lower tier local authorities (see RCLG 1960: 59; Self 1962: 146). According to Rhodes (1970: 230-233), three arguments regarding the London government reform during the 1960s exist. First, the functional argument, where the debate was almost entirely concerned with the effective provision of services. On the one hand, there was the need for effective delivery of services such as traffic, planning, and overspill housing in Greater London. On the other hand, it was argued that the existing system was providing a good service in areas such as education and child care and thus it would be a mistake to destroy it. Second, the argument about the value of local self-government which meant reinvigorating local government in London. This idea was seen not in terms of the size of authority (as was implicitly argued by the Herbert Commission) but in the existence of distinct communities which may be as large as a county or as small as a parish. Thus, from that point of view, the creation of the GLC and the London Boroughs represented a strengthening of local government. Third, the political argument, where all parties involved - i.e. local authorities, professional bodies, the Conservative and Labour parties - each sought certain specific aims. For this author, more than an argument in itself, these aims conformed to the underlying motives for reform; the main political argument was that the Labour Party accuse the Conservatives of 'gerrymandering and seeking a purely party political advantage in destroying the LCC' (Rhodes 1970: 232).

The two basic principles in the Herbert Report portray how primary concerns and the affirmation of the traditional values of local self-government - i.e. pluralism, participation and efficiency in the delivery of local services - have been influential in Britain. The debate during and after the period of the Herbert Report focused on the structural and institutional changes that would increase the efficiency of public service delivery as the value local government was to pursue. The emphasis placed on the traditional value of 'local democracy' by this, and subsequent reports, led to a series of criticisms as local government was pictured as democratic but inefficient (see, for example, Cochrane 1993: 116; Dearlove 1979: 56-59; Dunleavy 1980: 7-8).

The central government did not fully endorse all the proposals made by the Herbert Commission. While at the upper tier a Greater London Council was set up in 1965, at the lower, the Government established 32 boroughs (plus the City of London) with an average population of about 250 000 rather than the 52 authorities with a population of 150 000 envisaged by the Herbert Report. The newly-created GLC was responsible for services such as strategic planning, housing, the Fire Brigade and major roads, taking control of London Transport in 1970. The new boroughs were to provide social services, housing, local roads, libraries, recreation and parks. The new two-tier system of government (which came into existence in April 1965 after the enactment of the London Government Act in 1963) allowed central and local authorities' participation (at the upper and lower tiers) regarding diverse air pollution functions. As seen in Table 4.2, few years before the GLC was abolished, responsibilities for air pollution were spread across all tiers of government.

According to Barlow (1991: 87-88), after 1965, the system of government in London went through several phases: from an initial period of adjustment to the new structure to the questioning of the role of the GLC and its abolition in 1985.[2] This author identifies three main problems that influenced the way in which the new structure operated: the weaknesses of the reform, the changing nature of metropolitan issues, and the course of events in London's politics (see also Flynn et al 1985; Rhodes 1972; Sharpe 1995c; Young and Garside 1982). First, the weaknesses of the structure are related to the functional allocation between the two tiers of government and the nature of the relationship between the tiers. One weakness refers to the fact that while the boroughs were given responsibility for a number of important functions that were familiar to the local government system and relatively easily coordinated, the GLC was given functions that were novel and difficult to integrate - e.g. planning. A second weakness involved concurrent functions between the two tiers of government. While some responsibilities were clearly spelled out in the London Government Act, others were vague. For instance, services that could be easily divided were refuse (disposal and collection) and drainage (main and local). Services that were more complex included planning, housing and transport. As it was not clear where responsibility lay for this last set of services, the result was that in some areas performance suffered as consultation and cooperation were not effective. One last weakness was the lack of subordination from one level of government to the other as the rationale of the reform was to create separate and distinct types of local government unit. Thus the boroughs tended to undermine the authority of the GLC and reduce its performance and efficiency as a strategic body, again, particularly in services such as transport, housing and planning.

Table 4.2 Government air pollution control responsibilities before the GLC's abolition

Central government	*Greater London Council*	*London Borough Councils*
General advice	General advice	General advice
Air pollution monitoring and measurement	Air pollution monitoring and measurement	Air pollution monitoring and measurement
Air pollution data bank	Air pollution data bank	
Enforcement powers		Enforcement powers
Legislation		
Standardisation of test methods		
Advice in form of Codes of Practice		
	Strategic planning and development control	
		Local planning and development control

Source: GLC (1983)

A second problem that influenced the new system of government was that during the late 1970s concern over metropolitan issues shifted from housing and planning towards the state of the metropolitan economy. Thus, such aspects as unemployment and inner city problems came to the top of the government's agenda. In addition to this, politics in London was a determinant problem for the existence of the new London's government structure (see Barlow 1991: 95-97). Changes in the balance of power in the GLC and several boroughs passing back and forth between Labour and Conservative provoked serious confrontations between the GLC and individual boroughs and between the GLC and the central government. Such confrontations caused a lack of continuity in policy and political direction, especially concerning the issue of housing. During the 1970s there was a growing movement among Conservatives to press for a review of London's government and there were calls for the removal of the GLC's housing powers and for its abolition. It was not until the late 1970s though, that the proposal for abolition was not only supported but also advocated by the central government. As Barlow (1991: 97) states, 'abolition of the GLC became an election promise and the focus of a personal crusade by the Prime Minister (Thatcher)'.

Sharpe (1995c: 120) identifies three main reasons outlined by the central government for abolishing the GLC. First, it was argued that the GLC had been an irresponsible financial overspender in terms of the government's stated policy of reducing total government expenditure as well as compared with the expenditure of local government as a whole. Second, that the primary function of the GLC, the planning role, was no longer relevant or necessary because it was the government's intention to downgrade that function and instead to let market forces largely determine land use. Third, that the case for establishing fully fledged, elected, tax-raising bodies was weak because they did not have enough functions to sustain themselves as viable local governments. Despite these technical reasons, there is wide consensus that the underlying motif for abolition was merely a political one (Cuchillo and Morata 1991: 250; Hebbert and Travers 1988: 1-3; Sharpe 1995c: 120-128). The Thatcher Government proceeded to abolish the GLC without any public consultation and despite increased public support for keeping it in operation:

> The proposed reform of local government is unlike previous ones; it has emerged without detailed analysis of the present system and the possible alternatives (Flynn et al 1985: ix).

So, what was gained in the past in terms of discussing the importance of the structure of government and distribution of functions regarding air pollution and related issues was ignored in the 1983 Government White Paper 'Streamlining the Cities' and in the 1985 Local Government Act for abolition of the GLC. The proposals to reorganise local government in

London did not address the issue of pollution control, i.e. failed to say which agency was going to be in charge of London-wide pollution prevention activities (see DoE 1983; GLC 1984 (31); LGA 1985).

The abolition of the GLC and the current system of local government in London is also very much the product of a mixture of liberal and conservative 'New Right' values adopted by the Thatcher Government (Stewart and Stoker 1995b: 192). The poll tax and the introduction of compulsory competitive tendering by local authorities are two good examples of liberal right ideas which imply decentralisation and considerable powers of local government. For the liberal new right, the value of allocative efficiency has no concern with redistribution for social justice. Instead, for the conservative new right, the value of allocative efficiency preempts those of liberty and participation. New right associated with conservatism, as opposed to liberalism, implies a centralist position where the central government specifies local functions, the constitutional position of local authorities, their taxation levels, standards and allocative efficiency in local service provision (see King 1995: 234-238).

The system of government in London is now best described as a simple, single tier of 33 unitary authorities: the London government is essentially a borough government (Hebbert 1991: 191). After the abolition of the GLC, the boroughs emerged as the main responsible authorities for the provision of local services in London becoming 'the sole remaining tier of representative, multipurpose local government within the capital' (Hebbert and Travers 1988: 109). The governance of London, though, includes several other bodies, such as central government departments, centrally-appointed bodies, private sector agencies, and borough joint committees, all serving a wide range of different functions (see Table 4.3).

In their book 'The New Government of London', Travers and Jones (1997: 10-28) provide a brief description of how London is currently governed. First, the boroughs have become the strong basic unit of London's elected local government. The three main ways in which the boroughs operate are either by carrying out their own specific functions, by jointly constituting a number of London-wide organisations (by statute or voluntarily), or by forming partnerships with other central bodies, community groups, or local businesses. Second, like most borough joint arrangements, government-appointed bodies provide many services London-wide. While enhancing central government intervention in London's local affairs, these centrally-controlled bodies are not locally accountable to the electorate and do not seem always to operate in a coordinated way. Third, the central government has many direct responsibilities for the government of London by means of appointing all the board members described earlier, and of determining the funding for the bodies concerned. The purpose of existing agencies within Whitehall is to coordinate the activities of central government as they affect the capital, to act as coordinators and sponsors of some policy areas in London (e.g. transport), and to bring together all relevant bodies in London to encourage them to operate in a more effective collective manner. Finally, there are a number of London-wide and more localised partnerships that are business oriented. While their income is largely derived from the private sector (although some public authorities may give support to individual initiatives), they have focused particularly on economic promotion and the encouragement of tourism in London. By operating and encouraging partnerships, these bodies allow diverse organisations to network and to build common agendas filling a gap in the London machinery, but without executive powers, therefore they do not substitute governmental action.

With the existence of a borough system of government, joined by a multiplicity of bodies (joint committees and partnerships) and an increased central government intervention achieved through institutional reforms, appointments and financial controls, the governance of London looks fairly fragmented. This situation has raised concerns about such issues as coordination, representation and local democratic accountability across London. When the GLC was abolished, many of the roles and responsibilities were redirected to either the boroughs or the central government, which has directly participated by means of appointing London-wide organisations. With more central control, the government has set out a framework for ensuring the implementation of its own objectives and the involvement of the private sector (Newman and Thornley 1997: 967-969). So, for example, in the case of air pollution, while from 1986 the boroughs continued to be the environmental health authorities (which includes pollution control responsibilities), other air pollution-related functions such as road traffic (all local roads) and planning (each borough is a local planning authority) were transferred from the GLC to the boroughs. Nonetheless, the central government (now through the DETR) has maintained some relevant involvement in both issues: local plans follow strategic guidance and 'considerable reserve powers' concerning roads and traffic exist (see Chapter 6).

Table 4.3 The governance of London: participants and functions

Main participants	*Selected agencies*	*Functions*
Central government	Cabinet sub-committee for London (all central departments)	Coordinate activities of central government affecting London
	Government Office for London (GOL)	Regional authority for London: contact point with boroughs, voluntary and private bodies
	Department of the Environment Transport and the Regions (DETR)	Strategic planning, housing, urban programme schemes, public transport and road systems
Centrally-appointed bodies	London Transport (LT)	Underground and buses
	Metropolitan Police	Police Authority
	London Docklands Development Corporation (LDDC)	Urban development scheme
	London Residuary Body (LRB)	Remaining assets from GLC and ILEA
Boroughs	Diverse local units	Arts, environmental health, libraries, housing, refuse collection, education, social services, planning, roads and street lighting, cemeteries
Borough joint bodies	Association of London Government (ALG)	Boroughs' representative in a range of London-wide bodies
	London Fire and Civil Defence Authority (LFCDA)	Firefighting and civil defence
	London Planning Advisory Committee (LPAC) and London Research Centre (LRC)	Planning and research functions London-wide
	South East Regional Planning Conference (SERPLAN)	Policy-monitoring, major planning and transport issues
Private sector (partnerships)	London First	Economic promotion and tourism
	London Pride Partnership	City development plan

Source: After Travers and Jones (1997)

At the local government level, while it has been argued that the boroughs include the best and the worst of the local authorities in Britain, most concerns relate not to the boroughs themselves, but to the ways in which they operate through diverse area-wide joint bodies within the overall system of government (Hebbert 1992: 142-143). Given the fact that the government of London is a borough government, boroughs and its joint committees play an important role, not only for coordination purposes, but toward citizen participation and representation. In terms of democratic structures, the London boroughs go into elections one year in four where new authorities (councillors) are elected for the whole city. While no borough is accountable to any other borough, they are to the local electorate. As shown in Table 4.4, while during the 1980s London was more or less politically balanced, it has not been so during the 1990s, where the Labour Party has gained political control of many more boroughs. While the boroughs are the directly-elected units of government in London, the borough joint bodies are regarded as indirectly-elected units since they are formed by the boroughs as constituent members. Democratic accountability within these bodies operates in distinct ways and dependant on the type of joint arrangement: some report to their own local authorities, others to central government.[3] So, for example, in the case of the London Fire and Civil Defence Authority (LFCDA), accountability of its members and other joint authorities is to their constituent local councils. The level of accountability might increase depending on the area - whether fire, civil defence and so on. While the general public has access to the Committee meetings and papers, it also reports to the Home Office and is scrutinised by the Local Government Ombudsman (Hebbert and Travers 1988: 72-77).

There are some joint arrangements in which the boroughs are obliged to be members without the possibility of opting out. An example of these bodies is, again, the LFCDA, a statutory joint authority with the right to precept on the boroughs for local tax resources and such functions as firefighting and civil defence. Members of this area-wide body must be elected local councillors; the LFCDA is accountable to their constituent local councils. There are others in which membership is voluntarily and a matter of political choice, thus lacking a complete coverage of London's interests and / or failing to be a real voice for Londoners as they have little or no power. Regarding the lack of complete coverage, the cases of the former London Boroughs Association (LBA) (Labour dominated) and the Association of London Authorities (ALA) (Conservative dominated) constituted a good example. While the former (regarded as a real forum for discussing London-wide problems) represented only 17 boroughs and the City of London, the latter represented the interests of the remaining 15 boroughs (Hebbert and Travers 1988). The amalgamation of these two rival local authority organisations in 1994 into the Association of Local Government (ALG) now covers the whole area of London. Nevertheless, while the ALG potentially represents a stronger voice area-wide in London it still remains a networking and lobbying body and thus may have influence, but limited power.

Table 4.4 Political control of London (before and after 1986; 1995)

	Before May 1986	*After May 1986*	*1995*
Conservative	18	13	5
Labour	13	14	21
Liberal Democrat (alliance)	1	3	3
No overall control	-	2	2
Independent	1	1	1
Lab-Lib Coalition	-	-	1

Source: Hebbert and Travers (1988); Newman and Thornley (1997)

The principle of indirect election and variable accountability of a borough's joint arrangement have been criticised for being cumbersome and, above all, locally undemocratic (see Flynn and Leach 1984: 43; Hebbert and Travers 1988: 74). While there does not seem to be any form of 'reporting back' from these bodies to citizens, the process of accountability diminishes even more due to the augmented existence of new networks where the public sector has a minority voice in a particular partnership. It is the private sector together with the central government who have been increasingly taking the lead in discussing London's governance. Such has been the case with the newly created (1996) Joint London Advisory Panel (JLAP) constituted by twelve Ministers of the Cabinet Sub Committee for London and the members of the London Pride Partnership (a business oriented body including such bodies as ALG and LPAC). While the JLAP meets in private with the London Pride Partnership, the line of accountability from these meeting back to ALG or

LPAC, and then to London boroughs and to the local electorates, has been 'tenuous' (see Newman and Thornley 1997: 983). Hence, any process of democratic accountability in the new partnerships or networks is to be considered indirect.

The London government obtains financial resources through central government grants, local taxation and from the private sector. While around 75% to 80% of the boroughs' budget is transferred from the central government, not all of them receive the same amount. All London boroughs, for example, receive a share of the Uniform Business Rate (UBR), a national tax levied by the central government and then reallocated to local authorities based on population rates. Another important way to raise resources in order to deliver local public services is the council tax, which has become the main source of local income for the boroughs. While financial resources have also been sought through the participation of the private sector, the London boroughs, like the rest of the local authorities in Britain have both their revenue and capital spending capped (Travers and Jones 1997: 37). Like the boroughs, London-wide joint bodies also have financial constraints from the centre: their budgets are allocated depending on political-party participation. Hence, these joint organisations also receive their resources from the boroughs themselves, for example, the LPAC, and / or from other minor sources, such as petroleum licences in the case of the LFCDA (Hebbert and Travers 1988).

In the years since 1986 - particularly during the 1990s - the debate has continued regarding the reorganisation of the local government system in London (see, for example, Travers et al 1991c; Travers and Glaister 1997). By way of illustration, Travers and Jones (1997: 10) distinguish six arguments that have been put forward for changes to the local government system in London. These arguments are: the need for a London-wide local authority and / or mayor; the lack of direct democratic accountability to Londoners for many public services provided; over-use of appointed boards and ad hoc bodies; poor standards of service provision in some boroughs; a lack of coordination of land use and transport provision; and the failure to tackle many key social and economic problems for lack of resources and/or political capacity.

Before the general election of May 1997, two distinct approaches to the future of the reform of London government existed. The first, put forward by the Conservatives, involved little or no change from the current arrangements. The second, supported by Labour and the Liberal

Democrats, implied the creation of a new London-wide elected authority (Conservative 1997; Labour 1997; Travers and Jones 1997: 32-34). After the Labour Party came to power in 1997, the new administration published 'New Leadership of London', a consultation paper setting out the government's proposals for an elected strategic authority for London: the Greater London Authority (GLA) made up by a directly elected mayor and a separately elected assembly of 24-32 members (DETR 1997b). According to the government, the creation of a GLA seeks to restore democratic city-wide Government to London, preserve and enhance London's competitiveness providing firm leadership, tackle London's problems (e.g. economic, social) and speak up for Londoners and their interests. While making repeated reference to the need of a strategic authority for London, the Green Paper included air quality management concerns - specifically regarding transport issues - as part of the arguments for creating a GLA:

> The GLA would be well placed...[to develop]...a London-wide strategy on air quality, designed to achieve national air quality targets within London, properly integrated with GLA strategies on the environment, transport and land-use planning (DETR 1997b: 29).

While the government has highlighted that the GLA would not be duplicating the work of local authorities (the latter would continue to have responsibility for implementing the resulting strategy locally), it has been argued that there is a need to take account of the boroughs' work and existing collaborative arrangements when managing air quality. According to Rydin (1997: 17-19) the GLA's role in air quality management should include responsibility for a London Air Quality Strategy (LAQS) and its integration with a London transport strategy. While continuing and consolidating the work of the LAQN monitoring in London, the GLA should also have the power to restrict the traffic in certain spatial areas as this cannot be effectively undertaken by individual boroughs. The latter could include both physical 'bans' on road traffic and road pricing. At the same time, the London boroughs should continue to have the power to declare local air quality management areas within their areas (taking account of the LAQS) and to monitor air quality across London.

At the time of writing, the British government published the White Paper 'A Mayor and Assembly for London' proposing the new arrangements for London (see DETR 1998). A new GLA would be made up by a directly elected mayor with powers regarding transport, economic development, strategic planning, police, fire service, and the environment (including air quality management), and by a separately elected assembly (formed by 25 members) which would act as a check and balance body with wide ranging scrutiny powers to investigate important issues on behalf of Londoners and to call the mayor to account. Londoners will vote on the government proposals in a referendum on May 7 this year.

Mexico City: a lifetime under federalism

Since Mexico became a Federal Republic in 1824, the boundaries and structures of the current government of Mexico City have been largely determined by two interlinked aspects.[4] First, by the co-existence of the Federal District in Mexico City's physical limits and beyond them (see Tena Ramírez 1970: 312-313). Second, by the resistance of the central government to reform such a special political entity - particularly during the twentieth century (see Ward 1990: 85). While during pre-hispanic and colonial times Mexico City represented the most important political and economic urban centre, the decision to place the Federal District - seat of the federal powers - in the core of Mexico City, reinforced its socio-economic and political position as the national's capital city. It is precisely because Mexico City continued to be the political and economic centre of the nation that:

> Mexico's leaders have denied the capital's local residents their own independent structures for democratic participation [preventing] residents with neighbourhood or other more parochial concerns from interfering with urban administrative goals and national development plans. This, in turn, has meant that until recently, Mexico City's populations have been forced to use national political structures to express local concerns (Davis 1994: 5).

Major accounts of the history of the city capital have not only concentrated principally on the origins and evolution of the Federal District, but in doing so, they have often reviewed the extension of its boundaries, its territorial location, and the political rights of its inhabitants (see, for example, Burgoa 1985; Tena Ramírez 1970; Zavala 1992). From 1824 to 1854, the territorial extension of the DF covered an area of 8.8 sq.

kms (a radius of five miles around the central square of the city). These boundaries started to expand in 1854 as new surrounding local units *'distritos'* were incorporated. The last major boundary reorganisation that took place during the last part of the nineteenth century came about due to a neighbourhood petition in 1880 on extending the limits of the capital. This petition was based upon the fact that the city had expanded and thus the services provided by local authorities within those extended territorial limits were not enough (see DDF 1992: 53). It was not until 1898 that the definite boundaries of the DF (as they exist now) were established covering an area of 1 499 sq. kms. (Gamboa de Buen 1994: 43). By the end of the last century, Mexico City was confined to what it is still referred to as the 'First Quarter' - *'Primer Cuadro'* - within the Federal District boundaries covering an area of 20 sq. kms. (Ward 1990: 35).

The increase of the physical limits of the DF however, ignored the federal principle that the seat of the federal powers should be placed in a small territory - as happened in 1790 with the creation of Washington D.C., capital of the USA (see Burgoa: 1985: 913). According to Tena Ramírez (1970: 313), while increasing Mexico City's socio-economic and political hegemony over the whole nation, the DF's territorial expansions of 1857 and 1898 went beyond such a federal rationale which was originally adopted in the early years of independence with the 1824 Federal Constitution. The fact that the DF has been seen not only as the seat of the federal powers explains why such a special entity has had a changing and diverse politico-administrative status over the past two centuries. Thus, this author (1970: 314-321) explains that sometimes the DF has been a variant of any other Mexican federated state with the possibility of self-regulating its own governmental structures, and others a unit whose organisational structures and authorities were originally created by the federal government but in charge of local functions and with its own local judiciary tribunal.

The issue of the location of the DF has been more prominent than that of its physical boundaries (see, for example, Burgoa 1985: 916). The discussion that took place during the nineteenth century on whether the Federal District should remain or not in Mexico City has overshadowed the debate on the governing structures, not only of Mexico City but in fact, of the Federal District. The system of government in Mexico City and the Federal District did have diverse structural arrangements in terms of size, boundaries, representatives, service functions, and so on. For example, from 1824 to 1835, in the same territory both the DF and

Mexico City's local unit of government existed: the *'ayuntamiento'*. The *'ayuntamiento'* was responsible for services such as health, police and administration but it was accountable to the Governor of the DF. In 1829, the *'ayuntamiento'* divided the Municipality of Mexico into diverse politically fragmented areas known as *'cuarteles mayores'* and *'cuarteles menores'* (see DDF 1992: 21-22; Gamboa de Buen 1994: 42).

From 1846 to 1857, when the federal system was re-established in Mexico, the DF (renamed as the District of Mexico in 1847) co-existed once again with the then local unit of government in Mexico City: the municipality. The DF was divided up into 8 central units - *'prefecturas'* - and 3 foreign units - *'prefecturas foráneas'* (see DDF 1992: 33-34). The new 1857 Constitution reassured that the seat of the federal powers would remain in Mexico City and confirmed the territorial extension of the DF (Gamboa de Buen 1994: 42-43). Again, the debate on reorganising local government at that time focused on whether the DF should go to another city or not. Once the French Intervention (1864-1867) concluded, the municipal system was reinstated in Mexico City and thus the possibility for the DF's inhabitants to elect its local government authorities (Burgoa 1985: 918). Overall, the basic framework of government throughout that century was the municipality despite numerous changes of regime and patterns of state and local government which shifted back and forth between centralisation and decentralisation. For strictly local matters, Mexico City and the other municipalities in the district retained their local units of government: the *'ayuntamientos'* (see Fried 1972: 653).

The debate on reforming local government in Mexico City during the twentieth century has focused on two main issues. First, the political legitimacy of Mexico City's government which addresses two distinct aspects: the co-existence of the federal and the local governments in the same territory, and the political rights of its inhabitants (i.e. direct elections of local officers after the municipality was abolished in 1928). Second, the delivery of public services which includes a review of structural issues such as politico-administrative arrangements, intra- and inter-relationships with other public and private bodies, and the coordination, efficiency and effectiveness of policy implementation and enforcement (see Burgoa 1985: 913-936; Díaz Alfaro 1992: 198-199; Fried 1972: 684-686; Nava Escudero 1992: 111-116; Tena Ramírez 1970: 311-325; Ward 1990: 85-91). Throughout, the discussion on reorganising the structure of local government in Mexico City has concentrated on issues of democracy and political rights and thus has overshadowed those related to city administration (Díaz Alfaro 1992: 198-199). Mexico City is not an exception to the rule that in most developing countries unrepresentative and (usually) repressive government structures are stressed more than efficiency and effectiveness within local government (Akin Aina et al 1991: 4-6).

After the Mexican Revolution ended (1910-1917), the Federal Constitution of 1917 specified the creation of a government unit to run the DF; it was not until 1928 however, that a major reorganisation took place in the capital. First, the DF emerged as a special political entity within the Federation and was divided into a central department (City Hall) comprising a number of quarters and 13 delegated units *('delegaciones').* Second, the municipal system of government in the Federal District and the local unit of government of Mexico City - the *'ayuntamiento'* - were abolished. Third, the newly created department and delegated units took over the government of Mexico City and the rest of the municipalities within the territory of the DF, respectively. The non-elected central government department - called the Federal District Department (DDF) in 1941 - acted until 1997 as the area-wide authority for the DF. Fourth, the President of Mexico became responsible for the administration of the government of the DF. A presidential appointee would act as a 'mayor' or head at the central department to assist the President in carrying out his/her duties concerning the DF's local affairs. Finally, within the structural arrangement of the central department, a neighbourhood or consultative council *('Consejo Consultivo')* would serve as a representative mechanism for all concerns and demands of the inhabitants of the Federal District and with the official purpose of aiding Mexico City's mayor in governing the capital. This organisation however, would only give recommendations without any executive or legislative powers (Burgoa 1985: 921-922; DDF 1992: 61; Ward 1990: 73).

According to Ward (1990: 73) two main reasons for reforming the DF in 1928 existed. First, the municipalities within the DF were in major financial difficulty and incapable of delivering essential services. Second, the intense political in-fighting among political parties was taking place through the local government unit in Mexico City, the *'ayuntamiento'.* The reform would achieve two diverse things: to marginalise the then Labour Party's strength in the municipalities and to create a less anarchic structure capable of improving city administration. In the same vain, Davis (1994: 23) explains that the administration, construction and revival of Mexico City after the revolutionary period required 'fancy footwork' and 'crafty political alliance-making'. The reform came about due to the ambition of ex-president of Mexico, Alvaro Obregón (1920-1924) who struggled to regain the presidency in 1928. During the late 1920s, one of the main opponents to Obregón's intentions was an organisation constituted by labour unions - *'Confederación Regional de Obreros Mexicanos'* (CROM) - who had political control over Mexico City's municipalities, the local police and was an ally of his contender, P. Elías Calles. While the latter constituted a real challenge to Obregón's candidacy, the constant conflicts between the CROM and other labour unions, between different municipalities over diverse political ideologies, and between the local police and striking workers, were posing a threat to the efficient delivery of services and living conditions in the capital. This 'chaotic' situation led Obregón to urge a reform of the government system in Mexico City:

> In an effort to restore order to the capital city and limit the CROM's capacity to damage both local politics and the national labour movement, Obregón and his political allies successfully introduced an initiative in the National Congress in July 1928 to abolish the system of popularly elected municipal rule in the nation's capital (Davis 1994: 61).

Although Obregón was assassinated shortly after the reform, and P. Elías Calles came into power in 1928, the non-democratic system of local government in Mexico City prevailed until the early 1990s. In 1941, the DF was territorially organised within the same 1898 boundaries. It was legally ascertained that the DF was constituted by Mexico City (formed by 4 *'cuarteles'* and considered as a politico-administrative unit) and 12 delegated areas *('delegaciones')* (see DDF 1992: 61). Between 1940 and 1980, the built-up area in the DF grew 7 times from aprox. 99.2 sq. kms. to 750 sq. kms, stretching beyond its limits into the territory of the surrounding federated state EdoMex (Gamboa de Buen 1994: 44-45; Icazuriaga Montes 1992: 29-30; Ward 1990: 40). In spite of this increase, the system of local government remained the same as that of 1928.

Once again, in 1970 under the Organic Law of the Federal District Department *Ley Orgánica del Departamento del Distrito Federal* (LODDF), some re-organisation took place and Mexico City as a politico-administrative unit disappeared and was replaced by four more delegated areas. From 1970 to 1997, the Federal District was made up by a central department or City Hall (the DDF) and by 16 politico-administrative units or *delegaciones* altogether under direct control of the DDF (DDF 1992: 61). During this time, the mayor and the heads or local mayors of the *delegaciones* continued to be appointed by the President, as from 1928. In practice though, the local mayors were usually nominated by the mayor for confirmation by the President (Ward 1990: 73).

From 1970 to 1987, many formulas were introduced in order to expand the scope of citizen representation. Thus, for example, in 1970, the *'Consejo Consultivo'*, instead of being made up by different labour or civic associations, was formed by diverse groups of neighbourhoods called *'Juntas de Vecinos'*. The mere recommendatory functions that this body originally had continued without any granted executive or legislative powers (DDF 1992: 62-63). The long life of the Consultative Council (1928-95) served as an important body through which the *Partido Revolucionario Institucional* (PRI) assured a relatively stable one-party rule in Mexico City for several decades.[5] The establishment of the *'Junta de Vecinos'* further added to political and institutional fragmentation of local structures for political participation, rather than remedying the problem of lack of democracy in the capital (Davis 1994: 216). The need to reform the non-democratic system of Mexico City governance led to the creation in 1988 of a local assembly with responsibility for creating *'bandos'*, setting-up local regulations, monitoring local authorities and so on. This body - the Representatives Assembly of the Federal District *Asamblea de Representantes del Distrito Federal* (ARDF) - was set up as an elected party-political representatives assembly with limited functions: merely as a 'watchdog' for City Hall expenditures and policy (Ward 1990: 87).

After the fraudulent presidential elections of 1988, the pressure to democratise Mexico City grew considerably. Although city management issues started to be included in the debate for reforming local structures of government in Mexico City (see, for example, DDF 1993; Gamboa de Buen 1994: 158-159), it was the main concerns regarding democratic issues which prevailed on the governmental political agenda for change. The debate during the first half of the 1990s - commonly known as *'Reforma Política del Distrito Federal'* - focused on four main issues. First, direct elections of both the mayor and the local mayors or *delegados* (which also included devolution of power to their *'delegaciones')*. Second, further reform of the Representatives Assembly (ARDF), i.e. the creation of a proper local legislature or congress. Third, jurisdictional fragmentation of the DF by creating five more delegated units within its territory. Finally, whether or not Mexico City should become a 32nd federated state. Nevertheless, the only two issues that were finally addressed and included in the 1993 and 1996 reforms were the direct elections of the mayor and the creation of a local legislature (see DDF 1993; Gamboa de Buen 1994: 151-161; Ward 1998: 121-122).

While the issues of local accountability and social participation have been the hallmarks of Mexico City's government debate (see Díaz Díaz & Perló Cohen 1994: 72-73), air quality concerns have not awakened discussion on the need to reform the organisational structures of local government. So, for example, the 1990 air pollution strategy - PICCA - did not address the issue of carrying out structural changes in existing units of local government in Mexico City. However, the need to coordinate environmental action against pollution in the Metropolitan Zone of Mexico City led the Mexican government to create an inter-governmental agency in 1992: the CMPCCAVM (see Chapter 2). This metropolitan body[6] (now called CAM) is a voluntary and permanent association of governments (central and local) that is regularly convened to discuss and try to agree on solutions to air quality management issues and other environmental problems in the MZMC. The CAM is headed every two years by the Mayor of the DF, the Governor of the surrounding federated state EdoMex, and the Secretary of State for the Environment (SEMARNAP) under a rotation basis; an Advisory Council, which is integrated by the main environmental group leaders, universities and research centres, senators and deputies, members of the ALDF, and the private sector, provides technical and scientific assistance (see CMPCCAVM 1994b; DDF 1996). The idea of creating a single metropolitan authority for the whole built-up metropolitan zone (i.e. DF and conurbated municipalities) for managing air quality - and other city management issues - thus taking over CAM's responsibilities, has not become the focus of debate for reorganising local government in Mexico City (see Ward 1998: 282-283).

The local government in Mexico City - the Federal District Government *Gobierno del Distrito Federal* (GDF) - currently has three main local authorities (branches): the executive (Head of City Hall), the legislative (Legislative Assembly), and the judiciary (Tribunal of Justice). The system of government in Mexico City (executive branch) is best described as a highly centralised, city-wide and mayor-headed local authority. The City Hall of the Federal District Government - previously known as the Federal District Department (DDF) - consists of the Mayor's

Office and a large number of hierarchical subordinated local units. The organisational structures of the executive branch are divided up into three main areas (see Table 4.5). First, the Mayor's office and a number of local units which include secretariats *(secretarías generales),* the auditing office *(Contraloría General),* the administrative office *(Oficialía Mayor),* delegated units *(delegaciones),* and the Attorney Office *(Procuraduría General de Justicia del Distrito Federal).* Second, deconcentrated units which include, again, delegated units *(delegaciones),* administrative units, and other local bodies *(comisiones).* Finally, public corporations or parastatals *(entidades paraestatales),* which include decentralised units, public-private partnerships, and so on. Since December 1997, the GDF is no longer an ad hoc central government unit but a proper unit of local government (see Constitution 1997; LOAPDF 1994).

As the local authority in charge of the city, the mayor has his/her own office and responsibilities. Those functions that are allocated to his/her office and cannot be delegated or transferred to other local bodies include financial reports to the legislative body, appointments within the whole government of the DF, policy formulation and contracting out of public services, and so on (RIAPDF 1995). The majority of the public services, though, are delivered by the secretariats, and by other deconcentrated and parastatal bodies. The 16 delegated areas in which Mexico City is politically and territorially divided, have the same mayor's functions except those that have been exclusively allocated to him/her or to any other secretariat. The only distinctive characteristic of these units is that each *delegación* can only operate within its own jurisdiction. As seen in Table 4.5, an enormous amount of overlapping functions exist inside the structure of the GDF. In addition to this, the governance of Mexico City includes several other bodies, such as central government secretariats, centrally-appointed bodies (parastatals and deconcentrated units), private sector agencies, and metropolitan joint commissions, all serving a wide range of diverse functions (see Table 4.6).

The mayor or governor of the City Hall is the only directly elected authority within the executive branch: all other authorities within the hierarchical structure of the City Hall are appointed by and are accountable to the mayor. For the first time since 1928, direct elections for a Mexico City mayor (who had traditionally been appointed by the President) took place in July 1997.[7] The newly elected mayor, who came to power in December 1997, will exercise his/her executive powers for three years and will become accountable to the local electorate. In addition, the mayor has to report to a new local Legislative Assembly of the Federal District *Asamblea Legislativa del Distrito Federal* (ALDF) on the diverse activities carried out by the GDF. In the year 2000 direct elections will take place once again for a mayor and, for the first time, for all other local mayors, i.e. heads of the 16 delegated units or *delegaciones*.

Table 4.5 Organisational structure of Mexico City's Government (GDF)

Agencies	*Functions*
Centralised units:	
Mayor's Office	Policy formulation and contracting out of public services, budgets approval, all appointments within City Hall
Secretariats and Delegated units	Local government, urban, social and economic development, housing, environment, public works and services, education, health, finance, public transport and traffic management, police
Auditing Office	Auditing
Administrative Office	Personnel, technical equipment
Attorney General	Justice
Deconcentrated units:	
Delegated units *Delegaciones* and *comisiones*	Public services provision, licensing, cultural and civic affairs, legal services, parks, markets, cemeteries, waste collection, roads, libraries and museums, urban sub-plans, land property
Paraestatals:	
Decentralised units and Public/private partnerships	Electric trams, underground

Source: Various

Map 4.2 The Metropolitan Zone of Mexico City (MZMC)

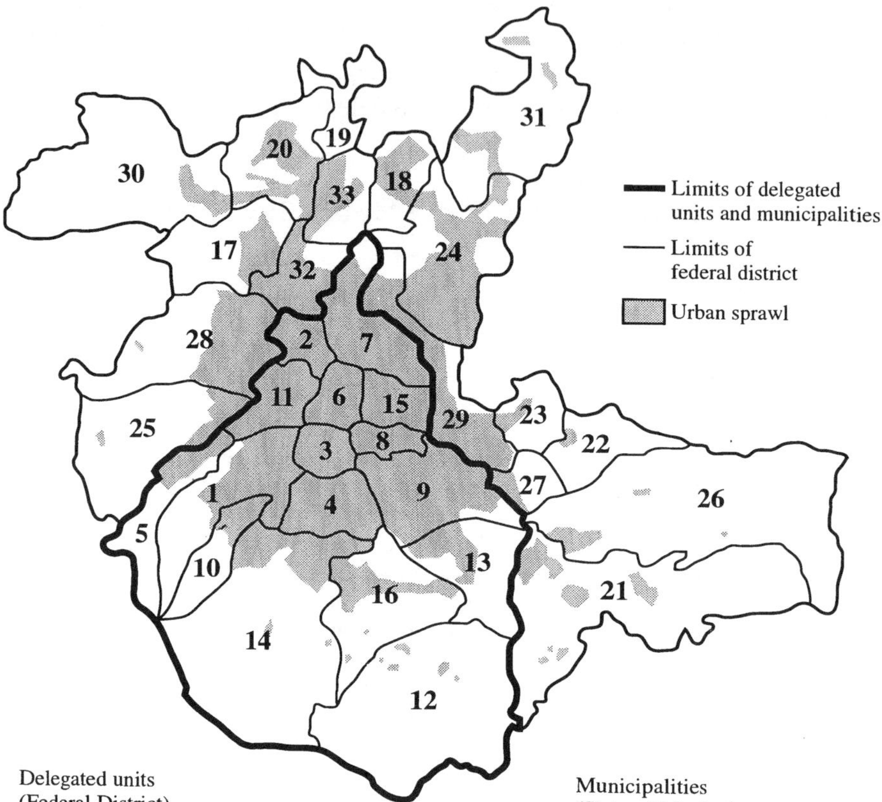

Delegated units
(Federal District)

1. Alvaro Obregón
2. Azcapotzalco
3. Benito Juárez
4. Cotoacán
5. Cuajimalpa
6. Cuauhtémoc
7. G. A. Madero
8. Iztacalco
9. Iztapalapa
10. Magdalena Contreras
11. Miguel Hidalgo
12. Milpa Alta
13. Tláhuac
14. Tlalpan
15. Venustiano Carranza
16. Xochimilco

Municipalities
(State of Mexico)

17. Atizapán de Zaragoza
18. Coacalco
19. Cuautitlán
20. Cuautitlán Izcalli
21. Chalco
22. Chicoloapan
23. Chimalhuacán
24. Ecatepec
25. Huixquilucan
26. Ixtapaluca
27. La Paz
28. Naucalpan de Juárez
29. Nezahualcóyoti
30. Nicolás Romero
31. Tecamac
32. Tlanepantla
33. Tultitlán

Table 4.6 The governance of Mexico City: participants and functions

Main participants	*Selected agencies*	*Functions*
Central government	Secretariat for the Environment (SEMARNAP)	Environmental norms and policy formulation, industrial inspection and control
	Secretariat for Energy (SE)	Energy saving programme
	Secretariat for Communications and Transport (SCT)	Transport emissions control (other than MZMC vehicles)
Centrally-appointed bodies	State-owned oil industry (PEMEX)	Fuels quality improvement, research
	State-owned electricity industry (CFE)	Electricity (thermoelectric plants)
Federal District Government (GDF)	Mayor's Office and diverse local units	Public works, water and drainage, planning, medical services, sports, fiscal revenue, economic development, social welfare, transport, traffic management, police, cultural and civic affairs, parks, cemeteries, roads, street lighting, urban plans, land property, waste collection, legal services, environment
Metropolitan Joint Bodies	Metropolitan Commission for the Environment (CAM)	Intergovernmental panel for environmental policy formulation (particular focus on air and water pollution, contaminated land)
	Metropolitan Commission for Transport and Traffic Management (COMETRAVI)	Transport systems and traffic management
Private sector	Diverse bodies	Taxi and bus services

Source: Various

As part of the political reform process in Mexico City, the legislative body experienced major changes in 1993 and 1996 gaining considerable legislative powers and becoming not only a proper parliamentary body (in practice since 1993, and formally since 1996), but a strong political agency within the decision-making process in Mexico City. The newly created ALDF, which has become an important player in democratising Mexico City, serves now as a mechanism for representing citizen's interests and has proved to be more effective for the latter than the already extinct Consultative (1928-1995) and Citizen's (1995-1997) Councils. As shown in Table 4.7, while the first legislative congress (still called Representatives Assembly) was largely dominated by the central government long-standing political party PRI, in the 1997 elections the left-wing party PRD gained most of the seats.

Table 4.7 Elected members to Mexico City's legislative body by political party (1994 ARDF and 1997 ALDF)

	Seats 1994	*Seats 1997*
PRI (right)	38	11
PAN (right)	14	11
PRD (left)	10	38
PT (labour)	2	2
PVEM (green)	2	4

Source: El Nacional, December 14, 1994; Business Mexico, November 1997.

As with the London's government, Mexico City's government receives its financial resources mainly from the central government, local taxation, and more recently, from the private sector. Traditionally, it has been the central government that has given financial support by means of subsidies, for example, for public transport, the underground system, sewers, and water supply. Only around 20% of Mexico City's budget came from local taxes. Nevertheless, central government's economic assistance has been considerably reduced over recent years leaving Mexico City's government to continuously depend on both their own resources (tax levy) and on schemes for borrowing money from the private sector. While local taxation rates for important services or commercial transactions, such as water or house acquisitions, exist, land property is by far the most important source for fiscal revenue (see Gamboa de Buen 1994: 107-117). Mexico City's first Governor-Elect, Cuauhtémoc Cárdenas, has assured in a recent interview that his administration will seek not only to attract new investment from the private sector but to promote public-private partnerships in such areas as public transportation and environmental pollution (see Cabannes 1997: 26).

Conclusion

The first part of this chapter brought into a comparative framework the main local structural arrangements in London and Mexico City. While London local authorities have no constitutional status, Mexico City's are the creation of the Federal Constitution and are threefold: executive, legislative, judiciary. Each governmental system operates with different tiers of government. On the one hand, without a city-wide authority, the government of London is constituted by a lower-tier of local elected authorities (32 boroughs and the City of London); on the other hand, Mexico City has a highly centralised upper-tier or city-wide local authority (headed by a mayor: the executive branch) with no elected lower-tier authorities (see Table 4.8).

While identifying existing differences in each urban centre, the description and analysis of the debates and experiences on governmental reform of each system of local government in the second part of this chapter depict few similarities about the changes or reorientations in their organisational structures. First, local authorities in both urban centres are becoming not the main, but one of the many players of a wider process of metropolitan governance. That is to say, London and Mexico City's local authorities are operating in a complex and growing system of diverse units (central government agencies, area-wide joint bodies, and private sector agencies) all providing a wide range of urban services.

Second, the current British and Mexican national governments have increasingly endorsed, though for different reasons, the idea to create more democratic institutions at the local government level. Clearly, the intention for London and Mexico City is to develop some type or variant of a metro scheme as explained in Chapter 3. In the case of London, while keeping the lower-tier of 33 local authorities, the British government seeks to establish or restore a democratic upper-tier of government. The Government proposals are to set up a new Greater London Authority which will include a mayor and assembly, both directly elected. In the case of Mexico City, while huge steps have already been taken to establish a more democratic upper-tier of government (by means of directly electing the mayor from 1997 onwards), a lower-tier of elected local authorities (most likely the 16 current delegated units) will start to operate in the year 2000. The principal difference here, however, is that in London the new authority will cover the full metropolitan area, not just one-half of it as is the case for Mexico City (see also, Ward 1998: 283).

Table 4.8 Comparative aspects of local government in London and Mexico City

	London (as Greater London)	*Mexico City (as the Federal District)*
Area (sq. Km.)	1 578	1 499
Population (000s)	6 933	8 235
System of local government	- NO legislative body NO supreme court	Executive: Mayor Legislative: ALDF Judiciary: TSJDF
Tiers of local government	**Upper tier:** NO city-wide authority **Lower tier:** 33 local units (32 boroughs and the City of London)	**Upper tier:** Government of the DF headed by a Mayor **Lower tier:** 16 local units (16 units without municipal status)
Democratic system	**Direct elections:** All 33 local units (1 year in 4)	**Direct elections:** Mayor (1997-2000; then every 6 years)

Source: Various

Finally, both urban centres are undergoing a process of continuous change in the organisation of their structures of local government. This ongoing process can be identified by the changes in the labels and functions of diverse local units whenever a new elected government comes into power. It is possible to affirm that both systems of government are embedded in a process of creation, fusion, fission and abolition of diverse local agencies.

While the debate on the creation of a city-wide authority in London has been invigorated by air quality concerns, the discussion on reforming local government in Mexico City has not included those related to air pollution control. The following chapters aim at examining the viewpoints of central and local government authorities as well as all other key actors in London and Mexico City government's organisational response to air pollution control, respectively. By exploring the cases of air quality monitoring systems, public transport and traffic management, and the application of an emergency plan, the argument centres on the need for a variant of metropolitan government.

Notes

1 Some of the best accounts of the history of the LCC, its structure and functional responsibilities, party politics, debates and ideologies for government reorganisation are to be found in Gibbon and Bell (1939); Jackson (1965); Robson (1939); Saint (1989); and Young and Garside (1982).

2 A good description and analysis of the history, arrangements and functional distribution of the GLC and London boroughs until 1985 include Barlow (1991); Regan (1972); Rhodes (1970 and 1972); Ruck and Rhodes (1970); Sharpe (1995c); Smallwood (1965); Travers et al (1991a); Young and Garside (1982).

3 According to Flynn and Leach (1984: 1-13) there is a distinction between joint boards and joint committees. Joint boards are corporate bodies, created by order of a Minister, with independent financial powers including the power to borrow and obtain the money it needs from constituent authorities by means of precepts. Joint committees have no corporate status independent of their constituent authorities, and cannot hold property, borrow or precept. They are creatures of the authorities creating them and their constitution and powers are controlled by, and may be terminated by, the constituent authorities; expenses are defrayed amongst constituent authorities as agreed.

4 After obtaining independence from Spain in 1821, Mexico has been governed in diverse ways. While the federalist structure has been predominant, Mexico has been also ruled twice by a monarchy (first during 1821-1824, and then during the French Intervention 1864-1867), and by a centralist republic (see Kandell 1988).

5 For a good review of the importance of the Consultative Council in Mexico City, see Davis (1994).

6 The CMPCCAVM was preceded by an inter-governmental panel - *Secretariado Técnico Intergubernamental* (STI) - formed by diverse central and local government units; this panel, created during the Salinas administration, launched the 1990 PICCA (see STI 1990).

7 From 1928 to 1993 the Head of the City Hall was directly appointed by the President of the country. From 1993 to 1996, the President would appoint the Head or mayor from a member of the local ALDF, the Senate, or the Chamber of Deputies - this formula, though, was never implemented.

5 Comparing Air Quality Management Strategies in London and Mexico City: the Local Government Perspective

By studying past and present air pollution problems and air quality management strategies UNEP and WHO believe that many of the problems currently faced by megacities can be avoided by the megacities of the future.

WHO & UNEP (1992) in their report 'Urban Air Pollution in Megacities of the World'

Introduction

This chapter brings into a comparative framework the viewpoints of local authorities in London and Mexico City regarding air pollution control and local government structures. The data regarding this comparative survey is based upon 29 structured questionnaires (out of a total of 33) for the case of London and 16 (out of a total of 16) for the case of Mexico City. These questionnaires targeted local authorities with environmental or pollution responsibilities according to the politico-administrative unit organisation in each city; in other words, the interviews were carried out with one local authority in each borough or *delegación* (see Chapter 1). All percentages that are shown in this chapter are based upon the total population of interviewees in London and Mexico City: 33 and 16 respectively. The four questionnaires that were not completed in the case of London are treated as a 'no response' answer.

The main purpose of this chapter is to review the lessons that can be learned from each city according to their own particular system of local government and air quality management regimes in the light of the three approaches outlined in Chapter 3. First, it seeks to consider past and present experiences of hierarchical structures of local government in relation to the various aspects of an air quality management system. Second, it highlights relevant points to be made about air quality (as a public good and thus its externality effects within a fragmented context) according to the public choice model. Finally, it attempts to depict current attitudes and network experiences, joint working and a variety of stakeholders for managing air quality in connection with the local governance approach.

Air pollution as an urban environmental priority

The starting point of the survey carried out in London and Mexico City tried to identify whether environmental pollution concerns exist in these two urban centres and whether or not the problem of air pollution is regarded by local authorities as an urban environmental priority. According to the results of the structured questionnaires, more than 73% in London and 81% in Mexico City of the local authorities interviewed, responded that environmental pollution as a whole (air, water, noise, contaminated land) is an issue of concern. However, whereas in London the environmental priorities are in first place noise pollution (67%) and in second, air pollution (52%), in Mexico City the environmental priorities are first air pollution (57%) and second water pollution (44%). The less important issues are, consequently, water pollution and contaminated land in London, and noise pollution and contaminated land in Mexico City (see Figure 5.1). The latter shows that whereas local environmental pollution as a whole may be an issue of concern for cities of the North and the South, governmental priorities for solving similar environmental problems may differ according to scientific data and political and social perception.

While considered as an urban environmental priority, air pollution is seen by local authorities in London and Mexico City as a local issue. Indeed, when local authorities require information about the sources, type, or quantity of pollutants, they usually get in touch either with other local authorities or other local agencies, or with central government and sometimes private sector bodies. So, for example, local authorities in London go either to other boroughs (24%), the central government (27%), the agency NETCEN (21%) or local bodies such as LBA or ALA (21%). They only go a limited number of times to the EC (6%) or to UN organisations (3%). These figures clearly reveal the lack of a centre to which local authorities could go and obtain standardised information about air pollution issues. In the light of orthodox views, there is a problem of equity in the distribution of this service because some boroughs may get more and better information than others depending on which body they get in touch with. The disparities of borough data on diverse pollutants (sources, types, levels) that may be found across London increases the difficulty of determining the desired performance or tolerance levels (that can be obtained vis à vis standardised databanks) for the efficient provision of public goods as outlined by the public choice model. This is because fragmented information on air pollution does not ensure the development of effective standards of measurement which in turn are necessary to analyse any production process as well as compare different modes of production for their efficiency (see Ostrom 1961: 833).

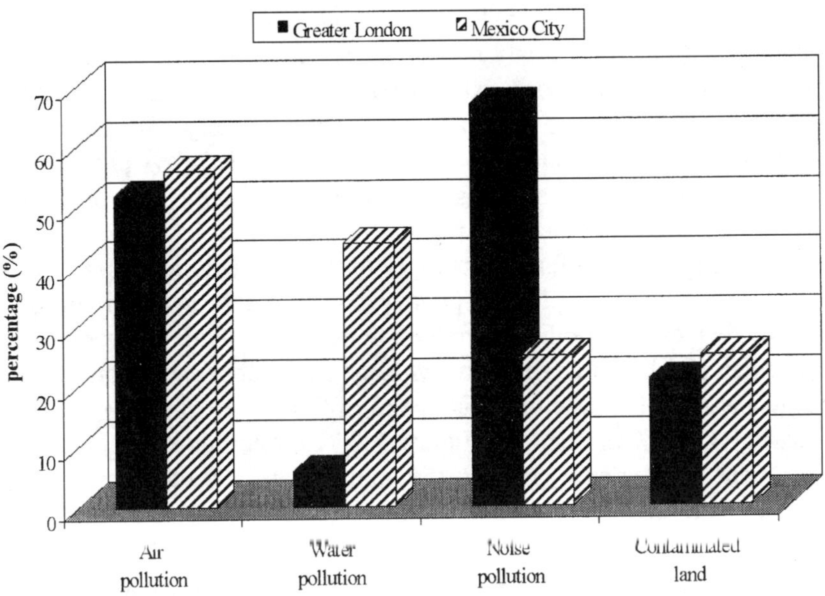

Source: Author's survey

Figure 5.1 Local authority views on different types of pollution in London and Mexico City

Additionally, such percentages for London show the existence of what the local governance approach identifies as the blurring of responsibilities between public and non-public sectors and power dependence issues. On the one hand, the presence of diverse stakeholders creates ambiguity and uncertainty about who is responsible for what and thus who should be accountable if the requested information is not appropriate or complete. On the other hand, the involvement of other sectors of society may bring enhanced knowledge, resources and technical efficiency. It seems clear, though, that the existing dependence of diverse boroughs on several other organisations for getting information on air pollution matters makes them one of the many participants without tangible leadership within the governance process in London.

By contrast, in the case of Mexico City local authorities go basically to the area-wide body DDF (87%), and rarely to other local authorities (19%) and the central government (13%). As with the case of London, they actually never go to regional bodies such as CEC (6%) or to UN organisations (0%). Information on diverse air pollution aspects can be obtained on a more standardised and equal basis which may enhance the possibility to measure and quantify data on the delivery of a public good. Nevertheless, a monopolistic position, as identified by public choice advocates, may derive from a situation whereby only one organisation provides all the relevant data on air pollution.

In addition to the fact that air pollution is regarded as a localised matter, local authorities do not always consider international air quality policy guidelines nor pollution reduction recommendations from regional bodies when tackling air pollution (with the noticeable exception of the EC and Agenda 21 in the case of London and of the Rio Declaration in the case of Mexico City). As seen in Figure 5.2, existing domestic air quality management strategies represent the main focus of attention for local authorities, i.e. the UK air quality management system in the case of London (82%),[1] and the PICCA in the case of Mexico City (57%). Although EC Directives are also a serious matter for London boroughs (82%), this may have more to do with reasons of legally-enforcing obligation than with a welcoming acceptance of EC recommendations. In the case of Mexico City, the fact that local authorities do not follow international guidelines is largely explained because *delegaciones* only represent the units through which the DDF's decisions are implemented, i.e. the operational administrative bodies.

Interestingly, as seen in Figure 5.3, strong consensus exists among local authorities that the most important aspect that needs to be revised in order to improve air quality is the issue of social attitudes, which includes such elements as education, awareness, and public participation (82% in London and 100% in Mexico City). However, while in London the issue of political will from central and local authorities is similarly very important

(82%), in Mexico City the second aspect that has to be taken into account is the need for more devolution of power to local authorities (94%). At the policy level, while in both urban centres current environmental policies also need to be addressed in order to improve air quality, in London local authorities seem to be less concerned about their implementation and enforcement (64%), than is the case in Mexico City (81%). The two aspects that are clearly identified as less important for local authorities in both cities are legal and taxation mechanisms and the economic model of development.

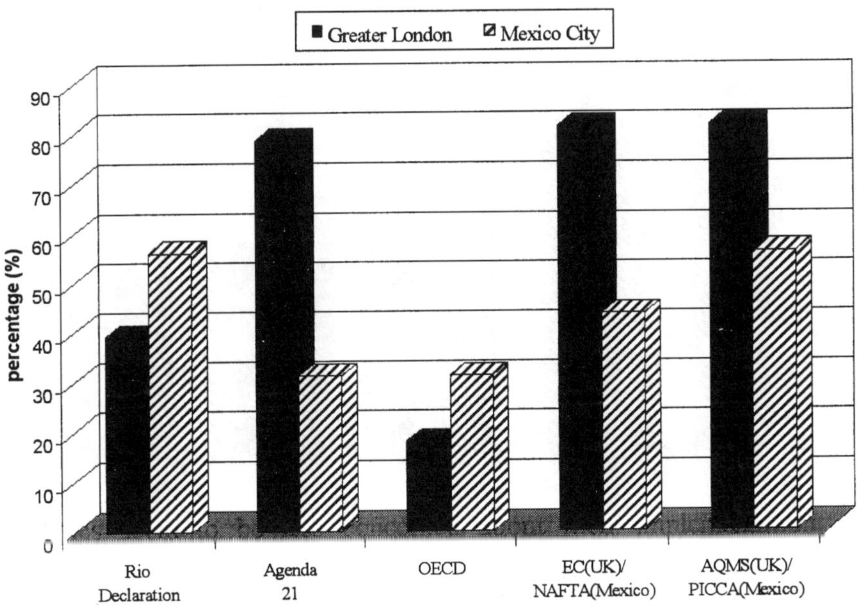

Source: Author's survey

Figure 5.2 Follow-up by local authorities in London and Mexico City to air pollution reduction recommendations

Results from Figure 5.3 also reveal a few concerns which are relevant to the local governance approach (see also Figure 5.4). In the case of London, while local authorities are concerned about issues of fragmentation (i.e. current local government organisation), they are more concerned about achieving air quality standards through the involvement of other actors of society, namely, citizens. Within the local governance approach, it is important to consider the participation of community groups or individuals through diverse local networks and various agencies. Here, concerns are not about pushing responsibilities on to the private sector or the citizen but about including other agencies or sectors such as voluntary groups, non-profits or non-governmental organisations (all recognised by the local governance approach) in managing air quality.

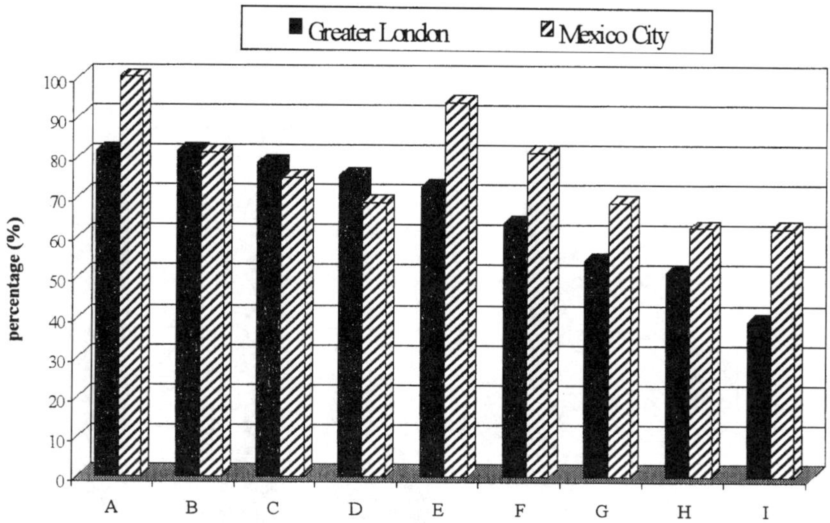

A Economic model of development
B Legal and taxation mechanisms
C Current local government organisation
D Policy implementation and enforcement
E Devolution of power to local authorities
F Local authorities finance
G Existing environmental policies
H Political will
I Social attitudes

Source: Author's survey

Figure 5.3 Local authority concerns on achieving air quality standards in London and Mexico City

Something similar can be said in connection with local governance assumptions in the case of Mexico City, where local authority concerns over the issue of social attitudes constitute one of the most important aspects to be addressed for achieving air quality standards. However, it is important to note that local authorities are also concerned about creating a more 'evenly balanced' structure of local government in Mexico City (as explained by the two-tier approach advocates) by demanding devolution of power to delegated units with more limited concerns on making the process of managing air quality more inclusive by increasing the participation of voluntary groups.

Local government organisation in London and Mexico City

As seen in Figures 5.4 and 5.5, while the most important factor that has limited local authorities accomplishing their air quality aims has been allocation of resources (67% in London and 56% in Mexico City) the latter is followed by two institutional constraints: central government intervention and the current organisational arrangements in both systems of local government.

From an economic point of view, one of the reasons why finance seemed to be a limitation for local authorities in London, rested on the fact that air quality was not a statutory duty and thus was not always a council priority, as highlighted by diverse London authorities.[2] This situation made securing additional funding for extension of air pollution activities almost impossible. As the local authority at Hounslow pointed out, air quality 'ha[d] to contend for finance with all other council functions including education, etc.'

Under public choice assumptions, this is a clear example of how the provision of a public good (i.e. air pollution) may create an externality. Without a duty to monitor or assess air pollution and without a central unit to financially shoulder local authorities across the whole metropolitan area, some boroughs in which emissions are produced may benefit from the actions taken by other surrounding boroughs who decide to allocate financial resources to improving air quality and thereby deal with the negative externality. With a duty to monitor and assess air quality (such powers were actually obtained through the 1995 EA) local authorities will now have to allocate resources to control air pollution. Yet, the boroughs where the effects of air pollution are tangible or where emissions are created (for example, with common heavy traffic jams or the existence of motorways), may have more incentive than others to carry out air quality management measures. More concerned and affected boroughs will only be able to internalise the externalities created by less concerned boroughs at a cost.

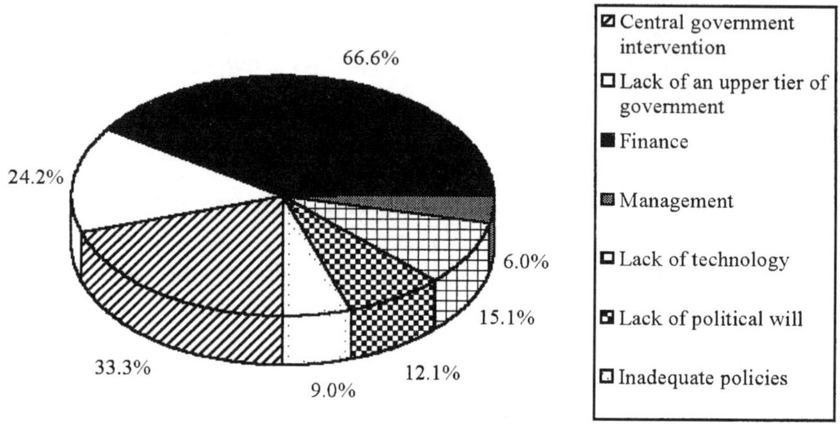

Source: Author's survey

Figure 5.4 London local authority main limitations for carrying out air quality aims

In the case of Mexico City, as *delegaciones* are only operational units and not 'proper' local authorities they usually have to follow what the Mayor's Office decides are the environmental priorities to work on each year and how much money is required for these. While *delegaciones* have little say from a financial resources perspective, the area-wide authority - the DDF - has the control of indirect consequences or externalities produced by different sources of air pollution across Mexico City. And while efforts to reduce air pollution may benefit surrounding municipalities of the federated state EdoMex, Mexico City inhabitants can be affected by the emissions produced within some of these surrounding municipalities (see Map 4.2). As seen in Chapter 7, the creation of a metropolitan council (i.e. CAM) for the whole metropolitan area has helped to coordinate environmental measures to tackle air pollution. Ideally, this body should encourage local authorities to internalise the externalities - positive and negative - for public 'good' such as air pollution which producers and consumers are unable or unwilling to internalise for themselves.

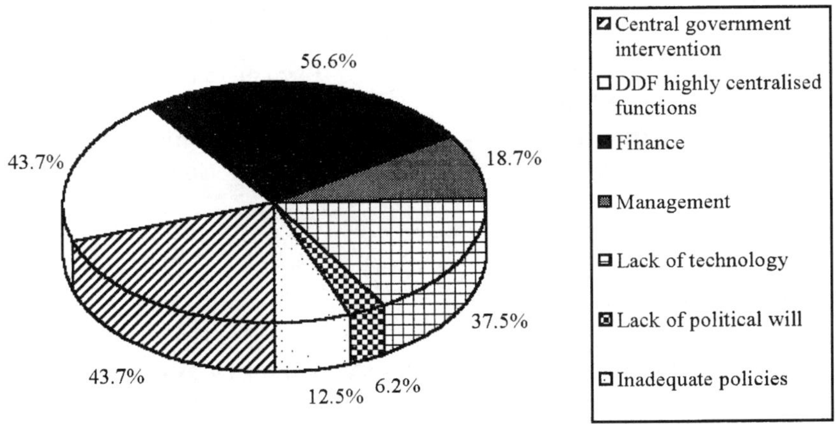

Source: Author's survey

Figure 5.5 Mexico City local authority main limitations for carrying out air quality aims

Central vs. local government intervention

The second greatest factor that has limited local authorities achieving their air quality aims is central government participation (33% in London and 44% in Mexico City). The results shown in Figure 5.4 and 5.5 demonstrate that although the issues of finance and local organisational arrangements have been very important in preventing local authorities carrying out their air quality responsibilities, they are not keen on heavy central government intervention regarding diverse air pollution issues. As seen in Table 5.1, local authorities believe that the central government is not the most appropriate agency for coordinating efforts between London and surrounding counties and between Mexico City and surrounding municipalities when dealing with air quality management issues, respectively (see also Figures 5.6 and 5.7 in the next section).

Table 5.1 Local authority preferences in London and Mexico City regarding central government intervention for coordinating air quality issues*

London	*Mexico City*
Local tier of government	Each of 16 *delegaciones*
Other specialised agency	Local tier of government
Central government	Central government
Each of 33 boroughs	Other specialised agency

*Preferences are ranked

Source: Author's survey

Concerning such issues as air quality monitoring systems, road traffic, and emergency plans, local authorities in both cities also prefer the participation of local government bodies to central government units or any other specialised agencies (see Table 5.2). In the case of London, the central government intervention option for coordinating and standardising air quality monitoring systems, road traffic, and an emergency plan, was put by the respondents either in second or third place as the 'more appropriate' agency for carrying out those functions. If the option of central government intervention appears in second place for the issues of road traffic and an eventual emergency plan, above a specialised agency, this may have to do with the fact that the latter would need considerable powers to change and implement road traffic programmes, and to ban or stop cars in the case of an eventual contingency plan. The situation in Mexico City looks clearer and more consistent: the option of central government intervention always appears in third place and below the upper-tier of government and the 16 delegated units options (see also Figures 5.7; 5.8; 5.9; 5.11; 5.12 and 5.13). It is important to highlight that in both urban centres the first option to coordinate monitoring systems, road traffic, and emergency plans, is an upper-tier of local government.

The results shown in Tables 5.1 and 5.2 reveal a mix of local authority concerns regarding the issue of central government intervention that is relevant to all three models explained in Chapter 3. There is a clear concern from local authorities for keeping air quality management regimes in local hands. From an orthodox perspective, (either a single consolidated unit or a two-tier formula) one of the reasons for creating a city-wide authority at metropolitan level is to avoid the responsibility of area-wide functions passing to central government. One of the main orthodox criticisms to the public choice approach is precisely that in a fragmented situation central government intervention will increase - although some public choice theorists would like to reduce the scope of central bureaucracy and arguments have been made for a more decentralised, democratic, and pluralist system within the process of governing urban centres.

Table 5.2 Local authority views in London and Mexico City on central government intervention for coordinating air quality monitoring systems, road traffic and emergency plans*

Air quality management issues	*London*	*Mexico City*
Air quality monitoring systems	Local tier of government Other specialised agency Central government Each of 33 boroughs	Local tier of government Each of 16 *delegaciones* Central government Other specialised agency
Road traffic	Local tier of government Central government Each of 33 boroughs -	Local tier of government Each of 16 *delegaciones* Central government Other specialised agency
Emergency plan	Local tier of government Central government Other specialised agency Each of 33 boroughs	Local tier of government Each of 16 *delegaciones* Central government Other specialised agency

*Preferences are ranked

Source: Author's survey

As explained in the following chapters, for example, much of the existence of heavy central government intervention and of diverse agencies for managing air quality in London is derived from the current institutional fragmentation at the local government level. Under traditional assumptions, obtaining governmental efficiency would imply creating a structure of government for maintaining and enhancing democratic accountability at the local level and for coordinating and standardising air quality issues (particularly regarding monitoring systems, road traffic and contingency plans), vis à vis the central government. Local authority preferences in London and Mexico City for such a metropolitan unit are linked to the critical issues or dilemmas outlined by the local governance approach regarding the need for an agency for coordination and leadership purposes. Although local authorities are also keen on including and encouraging the participation of new agencies (for example through self-governing networks) the outcome of the survey reveals that local authorities still wish to dominate in a governance relationship. Other specialised agencies do not seem the most appropriate units for coordinating air quality management issues.

Area-wide agency vs. fragmented units

As seen in Figures 5.4 and 5.5, the third most important factor that has hindered local authorities from carrying out their air quality management responsibilities is the way in which the system of local government in both centres is structurally organised (24% in London and 44% in Mexico City). The results of the survey show that in London the lack of a local tier of government for managing air quality is subjected to much debate among the respondents, in other words, there is no consensus on whether this has been a limitation or not. By contrast, in Mexico City the heavily centralised functions of the Federal District Department (DDF) are less controversial for the respondents and have represented a significant limitation. In spite of this, such limitations have not been a determinant issue that has impeded local authorities from doing their work or activities.

Local authority responses on which should be the appropriate agency for coordinating diverse air quality issues point in two clear directions in each city. In the case of London, the first clear point is that local authorities favour the idea of an area-wide agency as more appropriate for coordinating air quality management between London and surrounding counties, monitoring systems (together with standardisation), road traffic and contingency plans (see Figures 5.6; 5.7; 5.8 and 5.9). Within London's institutional setting, among the four given options, local authorities considered that the proper agency should be an elected body, i.e. a local tier of upper government such as the former Greater London Council (GLC) or proposed Greater London Authority (GLA). The second point is that the less favoured agency for coordinating air quality issues are fragmented units, that is to say, each of the 33 boroughs in London.

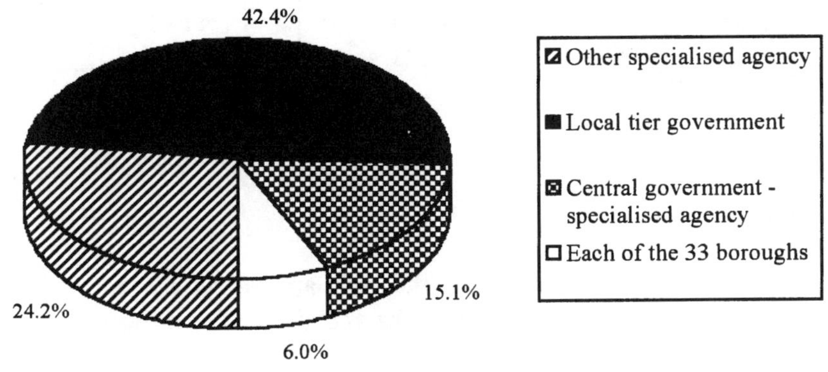

Source: Author's survey

Figure 5.6 Local authority views on the most appropriate agency to coordinate efforts between London and surrounding counties on air quality issues

First, regarding the issue of air quality monitoring systems, before the abolition of the Greater London Council a coordinating head in London existed that assisted local authorities on measuring, comparing and storing information on databanks on air quality trends (see Chapters 2 and 6). Due to the existence of such a body, which was in charge of collecting all air pollution data, it was possible to have an overall picture of Greater London's atmosphere. As seen in Figure 5.7, while most local authorities would like to have an upper tier of government to standardise and coordinate the results that they get when monitoring within their own boroughs, they believe that the second best option is a specialised agency. This is because local authorities tend to identify the South East Institute for Public Health (SEIPH) as the 'substitute' for what the former GLC was doing at the coordinating level for air pollution databanks (for details, see Chapter 6).

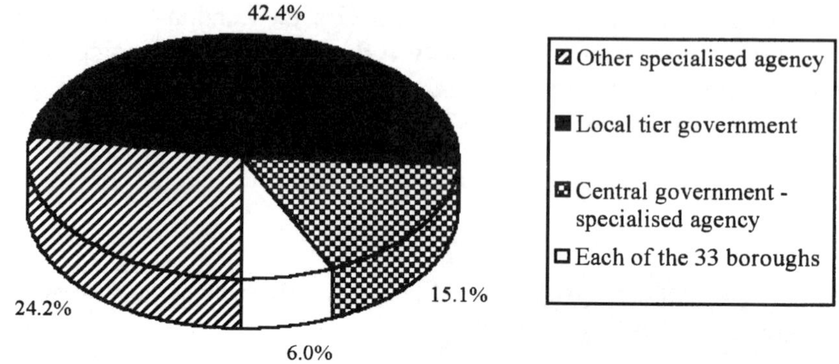

Source: Author's survey

Figure 5.7 Local authority views on the most appropriate agency to coordinate and standardise the air quality monitoring system in London

Second, local authorities are clearly in favour of an upper tier of government for coordinating road traffic issues (see Figure 5.8). Any other specialised agency is insufficient for this since what it is required is a head with enough powers to change things. The fact that the 'each of the 33 local units' option got a very low percentage in terms of local authority preferences, does not mean that they are not keen on getting involved in road traffic issues. On the contrary, local authorities in London have lobbied for a long time on issues such as testing and stopping vehicles. Although they do now have these two statutory powers, at the moment the survey was being carried out they lacked the powers to stop cars (see also Chapter 6).

Finally, regarding the coordination of an emergency air pollution plan, local authorities believe that this should be undertaken by an upper tier of government (see Figure 5.9). As with road traffic issues, a London-wide agency must have enough powers to change and implement such a plan. If a contingency programme is likely to contemplate restrictions in the use of cars, local authorities would certainly play a significant role in this. The responses by local authorities regarding an air pollution emergency plan have been given under the circumstances that there is not such a programme for London (see Chapter 6).

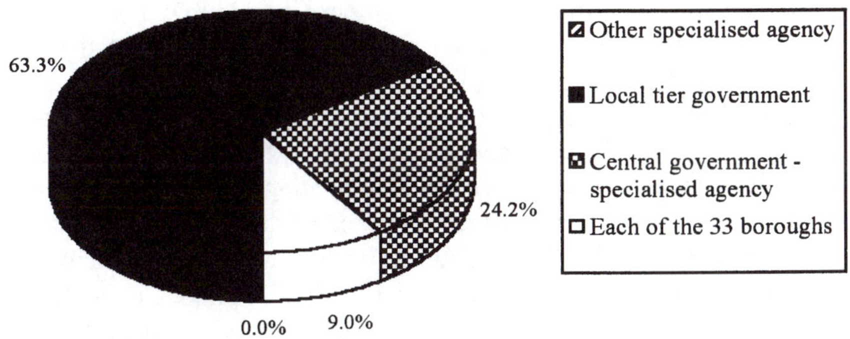

Source: Author's survey

Figure 5.8 Local authority views on the most appropriate agency to coordinate road traffic in London

The answers given by local authorities in London fit more readily into the 'hierarchical' and 'network' approaches (particularly for coordination purposes) than to the 'market' or public choice approach. Past and present experiences of London Boroughs working within a fragmented institutional setting seem to suggest that various political jurisdictions operating through contractual and/or cooperative undertakings to deliver public goods and services as proposed by public choice theorists do not constitute the most appropriate organisational arrangement for managing air quality. Rather, local authorities in London would prefer to work together with an upper tier of government and other participants (such as specialised agencies or central government units) in a more networking or partnership mode as outlined by a modified local governance approach. The outcome from Figures 5.6 to 5.8 reveals that within an interactive governing process for managing air quality (where several actors intervene), an area-wide authority is seen as the one organisation to dominate within the overall governance system. This qualifies the answers given in Tables 5.1 and 5.2.

In Mexico City, as with the case of London, local authority responses support first the idea of an area-wide agency for coordinating such issues as monitoring systems (together with standardisation), road traffic and contingency plans (see Figures 5.11; 5.12 and 5.13). Within Mexico City's institutional setting, among the four given options, local authorities believe that a local tier of government, such as former City Hall DDF should be the proper agency for this. The second point of relevance is that the second most appropriate agency for coordinating those three issues - unlike the case for London – was precisely the fragmented local units in Mexico City, that is to say, each of the 16 *delegaciones.* The latter, together with the fact that the first most appropriate agency for coordinating efforts between Mexico City and surrounding municipalities was each of the 16 delegated units (see Figure 5.10), reveals local authority desire to have wider participation in their dealings with air pollution. It is not that Mexico City's delegated authorities seek to set up a system of fragmented government but one in which they can jointly work with the area-wide body. All local authority responses implicitly acknowledged that the former DDF would remain the upper tier of government.

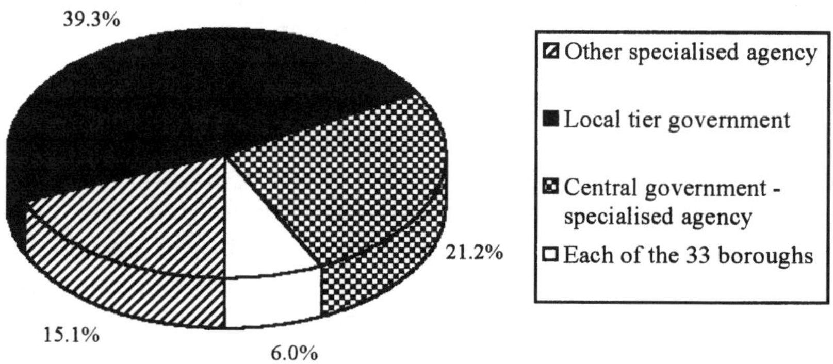

Source: Author's survey

Figure 5.9 Local authority views on the most appropriate agency to coordinate an emergency air pollution plan in London

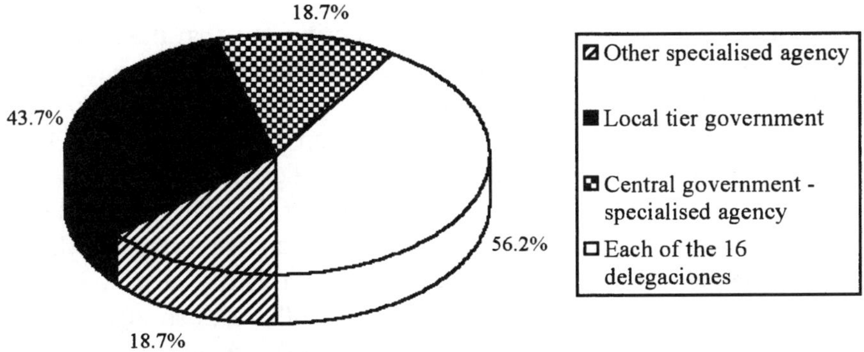

Source: Author's survey

Figure 5.10 Local authority views on the most appropriate agency to coordinate efforts between Mexico City and surrounding municipalities regarding air quality issues

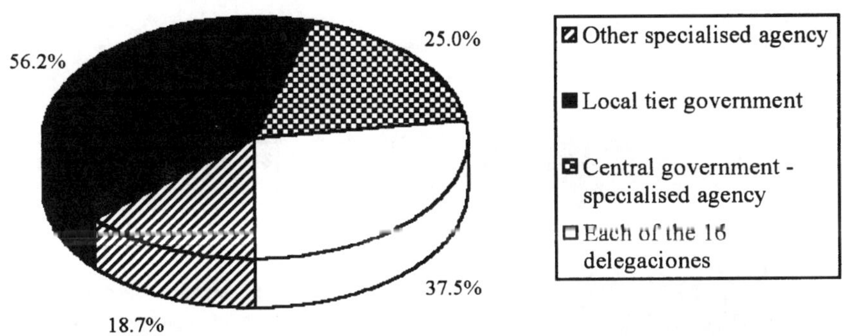

Source: Author's survey

Figure 5.11 Local authority views on the most appropriate agency to coordinate and standardise the air quality monitoring system in Mexico City

First, the case for air quality monitoring systems suggests that the agency for coordinating and standardising information on air pollution should be an upper tier of government and not a non-governmental area-wide agency (see Figure 5.11). Despite some criticisms that the media and the delegated units themselves have made regarding the monopolistic attitudes of the former DDF regarding monitoring system information (for example, not enough distribution of information to *delegaciones*), local authorities still prefer the option of an upper tier of government. The concern of local authorities is not so much whether the DDF should be the body for coordination purposes, but regarding access to air pollution databanks and general information (see also Chapter 7).

Second, as with the case of London, local authorities strongly believe that an upper tier of government (and not any other body) should be in charge of coordinating road traffic (see Figure 5.12). The reasons for this seem to be similar to those in the case of London: what is required for road traffic is a head with enough powers to change things. As the local authority in the delegated unit of Cuauhtémoc stated 'what is required is an entity with general powers so as to have proposals with integral solutions'. The results also show, once again, how much delegated authorities would like to have wider involvement on air pollution issues. The local authority at Tlalpan explained that the reason why *delegaciones* should intervene in road traffic issues is 'because of being more directly and closely involved with the problem'.

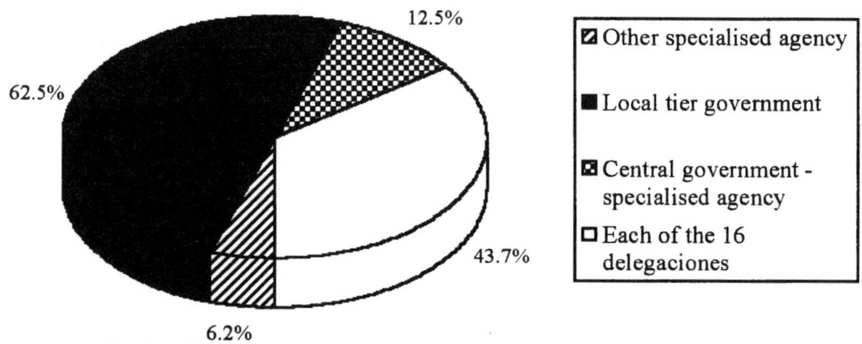

Source: Author's survey

Figure 5.12 Local authority views on the most appropriate agency to coordinate road traffic in Mexico City

Finally, results on the issue of the contingency plan suggest the need for a system of government with a head for coordination purposes and delegated authorities to carry out diverse measures of the plan (see Figure 5.13). Implementing a contingency plan requires an area-wide authority which would alert other bodies and coordinate and enforce several actions. Delegated authorities can also participate by taking quick action on informing people about the emergency situation in their own boundaries (e.g. hospitals, schools, etc.). As the local authority at Tláhuac stated on the implementation of the contingency plan: 'it has to be the local government of the DDF in conjunction with the 16 delegated units'.

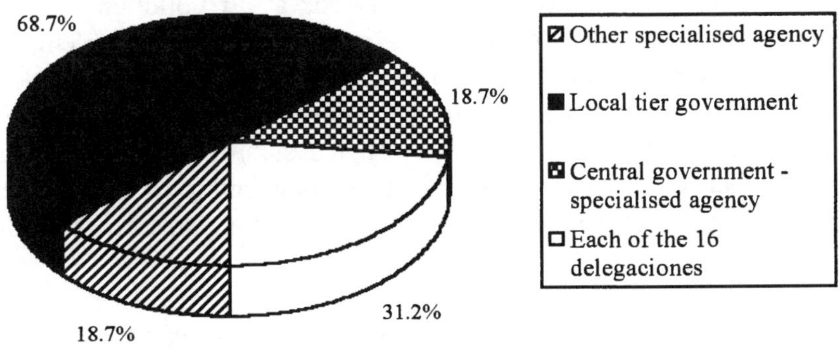

Source: Author's survey

Figure 5.13 Local authority views on the most appropriate agency to coordinate an emergency air pollution plan in Mexico City

The answers given by local authorities in Mexico City fit more with traditional orthodox assumptions (in particular the metro model) than with the public choice and local governance approaches. As seen from the results of Figures 5.9 to 5.13, although the participation of diverse stakeholders (particularly citizens) may be desired, the responsibility for managing air quality according to the respondents should rest upon local government units. Local authority desire (i.e. delegated units) to jointly participate with the DDF in air quality management issues is linked to the orthodox assumption of the two tier formula with a 'top-heavy' (or 'evenly-balanced' for some functions) type of metro authority (see Chapter 3). It is important to note however, that while central government units and specialised agencies got the lower percentages as the most appropriate units to coordinate diverse air quality issues, equal concern among all respondents exists to see them participating to a certain extent within the overall governance process.

Comparing air quality management systems

This section looks at the diverse ways in which local authorities in Greater London and Mexico City perceive and look at air pollution. The analysis concentrates on local authority views on the scale of the problem, the need for a contingency plan, the air quality situation (according to national and international guidelines and bands), the existing access to information and on diverse aspects on public transportation systems and traffic management. The last part of this section concludes by commenting on the relevant links of all these issues (policy measures) to the three models as outlined in Chapter 3.

Human health, flora and fauna, and the built environment

Local authority awareness of the risk or evidence that exists between exposure to any air pollutant and a respiratory disease or other illnesses in human beings is higher in London (58%) than in Mexico City (44%). However, the involvement of local authorities in programmes and/or research on air pollution and its effects on human health (London 33% and Mexico City 13%) is considerable low in both cities - particularly in the Mexican case.

In London, local authority responses regarding the awareness of a cause-effect relationship of air pollution may be scientific because some of the respondents (specifically Environmental Health Officers) have environmental science backgrounds. However, these responses included a wide range of answers such as 'anecdotal only', numerous references - too many to mention!', or even 'could write a book in answer to this question!'. Due to the length of responses, type of pollutants-consequences statements, and so on, it is not possible to transcribe all of them. Nevertheless, it is important to highlight that the two common characteristics given by those that perceived air pollution as a harmful issue were, first, that much of the awareness comes from academic related sources (e.g. medical journals; health articles); second, that the air pollutants identified are the typically urban pollutants such as NO_2, smoke, CO, SO_2, particulate matter, and the 'new' threatening O_3, PM_{10}, and benzene. The most common health consequences identified by London authorities are asthma, lung cancer, heart disease, bronchitis and asbestosis. Research that has been carried out by local authorities on air pollution and human health include, for example, the London Borough of Barnet on asthma in children and air pollution; the boroughs of Greenwich and Tower Hamlets (jointly with UMDS Guy's and St. Thomas's Medical and Dental School) on air pollution and respiratory health; and the City of Westminster (together with St. Mary's Hospital) on individual exposure.

In the case of Mexico City, awareness of air pollution impact on human health is relatively low compared to the high levels of pollution that the city commonly suffers. The latter may have to do with the fact that local authorities have limited access to DDF information on air pollution issues, and also because most of them do not have environmental science backgrounds. Most importantly, though, it may have to do with the fact that they are not commonly involved in any programme or research carried out by DDF authorities (or any other institution) on air pollution and human health impact. The most common consequences that were identified by the local authorities that were aware of this cause-effect relationship, include asthma, skin infections, bronchitis, heart disease, lung cancer, and diminished learning capacity due to levels of lead in blood. Local authorities from diverse delegated units have also identified the typical urban pollutants (e.g. lead, CO_2 SO_2 NO_2) and the 'new' threatening ones (e.g. ozone, benzene and some hydrocarbons) as the main pollutants that may cause harm to human beings. According to the survey, one of the few examples of a *delegación* taking part in air pollution and human health research, is the delegated unit of Coyoacán, which jointly participates in a National Autonomous University of Mexico (UNAM) research-led programme on ozone and illnesses in children.

Contingency plans

Whereas in London more than 70% of local authorities believe that an emergency plan is necessary in case air pollution levels exceed national or international standards for London, in Mexico City, 100% think that the emergency plan should continue to be implemented. Although the levels of some pollutants in London have gone beyond the permissible limits set out in WHO/EC standards, an emergency plan in this city does not seem to be as urgent as it is in the case of Mexico City, where the WHO/IMECA threshold limits have been regularly exceeded (see Chapter 2). As seen in Figure 5.14, among the three given options for an emergency plan, London authorities seem to prefer an all-year alert plan; nevertheless, there is not a common view on when should this plan take effect. On the other hand, in Mexico City, local authorities seem to be much more concerned about the implementation of the alert plan during wintertime.

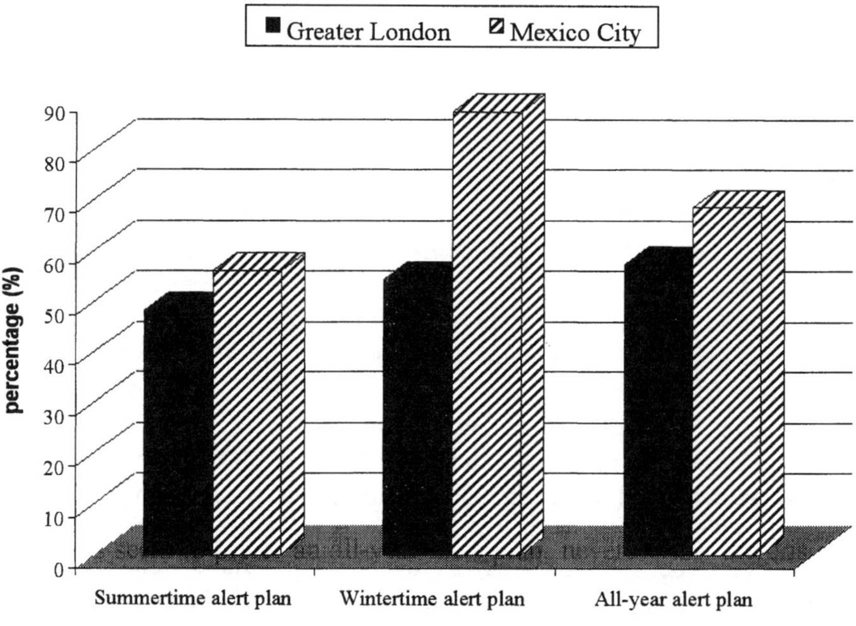

Source: Author's survey

Figure 5.14 Local authority concerns on the implementation of an emergency plan in London and Mexico City

Interestingly, although the so-called 'photochemical episodes' are more likely to occur in the summer, respondents in both cities are more concerned about wintertime pollution episodes than summertime ones. In the case of

London, this may have to do with the historical fact that severe pollution episodes have mainly occurred during the winter. In the case of Mexico City, while air quality worsens during the winter due to meteorological conditions, severe pollution episodes have occurred during all periods of the year - especially from March to May when mild-hot weather mixes during the early mornings and afternoons, wind-speeds are low, and there is no rain (for more details, see Chapter 2).

Bands and guidelines

In London, local authorities do not favour the current DETR's air quality criteria in which air pollution concentration is banded - i.e. 'very good', 'good', 'poor', and 'very poor'. The reasons given by the 76% of London authorities who responded in this way are that the bands are too wide and vague, imprecise and not strict. These have been also criticised because they do not account for mixtures (synergetic levels) and do not demonstrate the levels of risk to human health.

In Mexico City, the air quality criteria (IMECA) in which air pollution concentration is banded is one of much debate among local authorities: 44% in favour and 44% against. The authorities in favour pointed out that the Mexican index IMECA takes into consideration the realistic and particular characteristics Mexico City has and not the ideal ones. In fact, a local authority at the delegated unit in Azcapotzalco said that the criteria is the right one for Mexico City precisely because in order to have such criteria 'it is necessary to make an evaluation according to the physical characteristics of Mexico City'. The authorities against IMECA said that the bands are too wide and do not take into account all types of pollutants; furthermore, they are presented in such a way that the public do not understand them.

As seen in Figure 5.15, in London and Mexico City the WHO health-related guidelines are preferred by local authorities over the domestic guidelines provided by the British DETR and the Mexican government, respectively. However, in both urban centres the respondents prefer such domestic guidelines to the US Environment Protection Agency Standards. While quite important for London's local authorities, the EC Directive limit values are less appropriate guidelines to be followed for Mexico City.

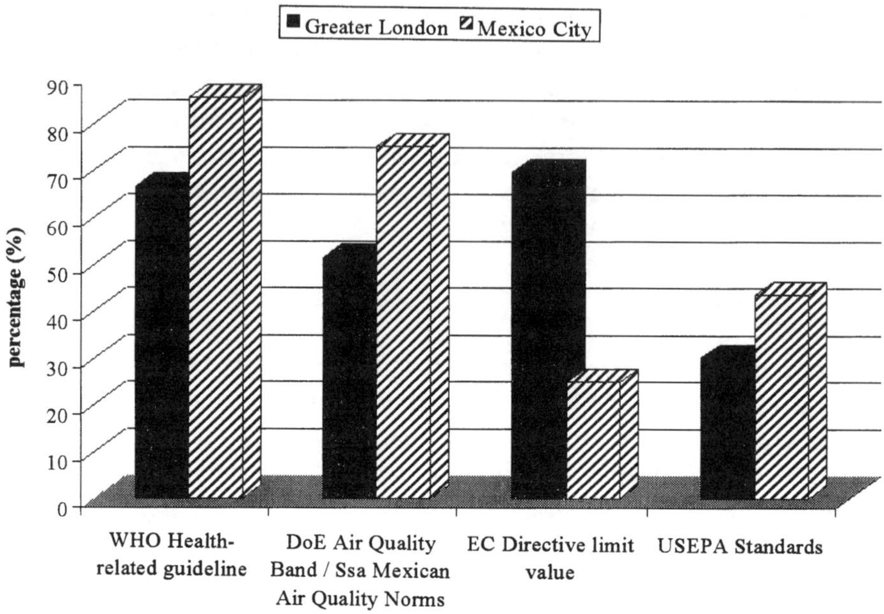

Source: Author's survey

Figure 5.15 Local authority preferences on air quality bands in London and Mexico City

Access to information

There is a strong tendency for local authorities in both cities to believe that the more effective channels to provide information about the air quality situation are through media - particularly TV and radio (see Table 5.3). These preferences may be linked to the fact that more people (especially vulnerable groups) have better and immediate contact with either TV or radio during the day and night than with any other channel of information. As seen in Table 5.4, the availability of information in London in this regard does not seem to be as good as it is in Mexico City.

The second most important channel to provide information according to the views of London and Mexico City's local authorities is the given option of signposts on main roadsides as this may help motorists to know about an emergency situation. Given the fact that road transport is the main polluter in London and Mexico City, this may be an appropriate way to alert drivers and ask them to make more 'rational use' of their vehicles. In Mexico City, signposts started to work at the beginning of the 1990s intended primarily to inform the population about the air quality situation. Some years on, their use of them has become less intensive since not all of them provide continuous information, and some are already broken.

Table 5.3 Local authority views on the most effective channel to provide information about London and Mexico City's air quality situation*

London	*Mexico City*
Media (TV, radio)	Media (TV, radio)
Signposts on main roadside	Signposts on main roadside
Local information (local Magazines, leaflets)	Newspapers
Tube stations	Tube stations
Newspapers	Local information (local Magazines, leaflets)

*Views are ranked

Source: Author's survey

Table 5.4 Local authority views (%) on availability of information regarding the air quality situation in London and Mexico City

	good	regular	bad
London			
Media (TV, radio)	9.0	48.0	30.0
Newspapers	3.0	42.0	42.0
Local information	6.0	18.0	58.0
Mexico City			
Media (TV, radio)	56.0	19.0	25.0
Newspapers	50.0	31.0	19.0
Local information	31.0	25.0	25.0

Source: Author's survey

Public transport and traffic management

According to the results of the survey, local authorities in London and Mexico City strongly believe that part of the solution to the problem of air pollution is the use of public transport. Whereas most authorities in London would like to encourage drivers to leave their cars at home and to shift to various modes of public transport, in Mexico City they prefer to have better public transport without so many limitations on private cars.

Table 5.5 Local authority views in London and Mexico City on the most important modes of transport to improve air quality

London	*Mexico City*
Railways	Underground
Buses	Environmentally-friendly cars
Underground	Electric buses
Cycling	Light railways *(tren ligero)*
Walking	Buses
Environmentally-friendly cars	Cycling
	Walking
	Railways
	Minibuses

*Views are ranked

Source: Author's survey

As seen in Table 5.5, the most important modes of transport that need to be encouraged in London in order to contribute to a better air quality standard are: railways (73%), buses and underground (64%), and cycling (61%). As explained by a local authority at the London Borough of Camden, one of the reasons why the option of more environmentally-friendly cars should not be encouraged as a mode of transport is because 'cleaner cars can provide a shorter term solution but longer term requires a shift away from cars'. In addition, local authorities suggested 'trams' and 'human powered rickshaw taxis' as alternative modes of transport.

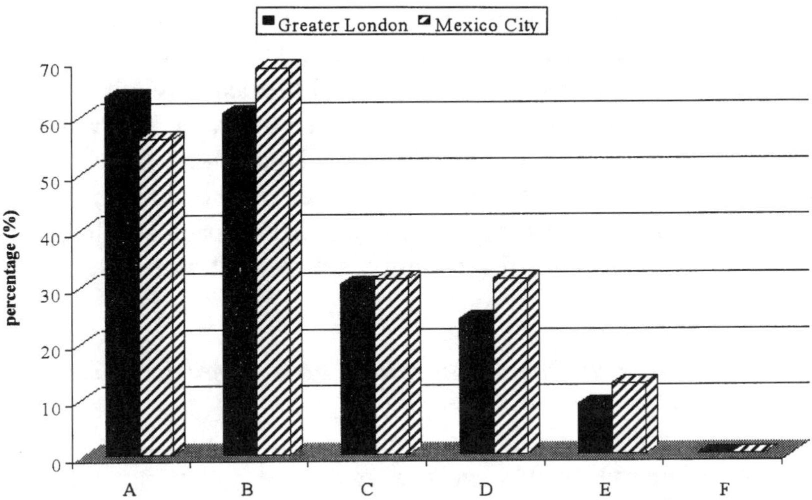

A	Consumerism and comfort	D	Safety reasons
B	Lack of public transport	E	Widening existing roads and building new ones
C	Social status	F	Less exposure to bad air quality

Source: Author's survey

Figure 5.16 Local authority views on why people are increasingly tending to make use of private cars in London and Mexico City

In the case of Mexico City, while local authorities believe that public transport (together with more environmentally-friendly cars) needs to be encouraged to improve air quality, not all modes of public transport are relevant. The most important of these are: underground (69%), electric buses (50%), and light railways (44%). As seen in Table 5.5, among the six given options of public transport local authorities believe that the minibuses (6%) - which account for the highest percentage of journeys in Mexico City by mode of transport (see Chapter 7) - are the least important. Cycling and walking do not seem viable options supported by local authorities; as one local authority in Tlalpan said, 'cycling and walking are not convenient in terms of distances and lack of security in the city'.

As shown in Figure 5.16, it is interesting to note that in both cities local authorities believe that people are increasingly making use of the car (main contributor to air pollution in both case-studies) because of lack of public transport (London 61% and Mexico City 69%) and because of consumerism and comfort (London 64% and Mexico City 56%). While the most controversial issue in London and Mexico City was the given option of social status (the results of the survey did not show any tendency on whether this has been a very important factor or not), local authorities in both cities do not consider that widening existing roads and building new ones, or less exposure to bad air quality, are relevant in making people use their cars.

Diverse policy measures and distinct local government approaches

From research presented in this section (Figures 5.14 to 5.16 and Tables 5.3 to 5.5) it is important to recognise that diverse policy measures may fit more readily with one or two of the three approaches outlined in Chapter III than others. Thus, there are some policy measures which may require an area-wide approach for a more efficient (where large scale economies can be exploited), and coordinated (where equal distribution of services among all political units is necessary), governmental response. Such are the cases of the elaboration and implementation of contingency plans and the elaboration of a master plan for enhancing public transport among diverse political units across a whole metropolitan area. The establishment of bands and guidelines may also remain an exclusive central/local government function. These issues can then be associated with the traditional orthodox model.

Arguing from the public choice approach, attention needs to be paid not only to the costs of production (efficiency) but to the transaction costs of the relationships of fragmented local units and how these might be reduced. Cases where externalities are not likely to occur may fit more to a fragmented system of government as explained by the 'market' approach. Some examples may include local bus services provision (in each jurisdiction) and research on human-health exposure to air pollutants in particular areas of a metropolitan complex where neither area-wide coordination nor cooperative agreements among local units may be required. In a few selected cases, there may be areas where the benefits of transaction costs greatly outweigh the additional transaction costs involved. As seen in this section, local authorities are working with other non-governmental agencies (such as universities, hospitals, private contractors) which may help local units to reduce the overall costs in the provision of a service.

However, with a metro-level public good such as air quality, the application of the public choice model is likely to remain limited.

Transaction costs need to be assessed, not only in relation to who produces and/or provides them (either a metro unit or multiple jurisdictions), but also in relation to the benefits derived from the relationships among diverse actors within a metropolitan area. There may be some policy measures where involvement of other stakeholders is required in the provision of certain services regardless of transaction costs; this can normatively be identified with the local governance approach. Some examples of this include access to and distribution of information (i.e. cost of the relationships of local authorities with media and newspapers) and delivery of diverse modes of metro-level public transport (e.g. underground, railways, trolleys, taxis). Here, focus needs to be put on networks and public/private partnerships for providing services.

Conclusion

The results of this survey are useful not only because they are representative of a population (i.e. local authorities in two urban centres) but because they provide a descriptive background for comparing the viewpoints of local authorities within each case-study and between the two chosen populations. While this comparable data fosters the exploration of diverse air quality management and local government issues as presented and explored in more detail in the following chapters, it serves as valuable empirical information to be evaluated and analysed in the light of the three approaches as outlined in Chapter 3.

The outcome of the survey with local authorities in London and Mexico City has suggested two main points regarding the organisational arrangements in each system of local government. First, most local authorities in both urban centres would like to see less central government intervention for managing air quality issues. While there is a similar view on more devolution of power to local authorities on air pollution matters, some central government intervention is still accepted. Second, there are strong pressures in both urban centres to reorganise local government into a system of an area-wide authority at the upper tier and participatory local authorities at the lower one. In the case of London, local authorities prefer working together with an upper tier of local government as a coordinating head for such issues as monitoring systems, road traffic, and emergency plan than separately with other boroughs or any other specialised area-wide agency either from the central government or not. Likewise, in Mexico City, local authorities believe that the most appropriate agency for coordinating air quality issues is an area-wide authority (with joint participation of delegated units) at the local government level.

The analysis and evaluation of the results of the survey within the framework of the 'hierarchical', 'market' and 'network' approaches strengthens the discussion on the advantages and disadvantages of each model for reorganising local government to improve air quality. In the case of London, while local authorities have raised concerns about the fragmented way in which they are currently operating, the extent to which London boroughs think they can solve major air pollution problems is revealed by the results in this survey. Local authorities believe however, that this participation needs to be done on a joint working basis with other agencies. Whereas local authorities favour the idea of a two-tier system as the most appropriate governmental arrangement for an efficient, equalised, and coordinated response to air quality management, they also believe that such environmental problems can be jointly solved with the participation of other sectors of society (either through networks or partnerships). As decentralised institutions in London (i.e. boroughs) are not keen on adopting an exclusive orthodox style of government in the form of a two-tier formula, the participation of local government units needs to be strengthened and enhanced within the overall governing process. The latter fits more readily with a modification of the traditional 'hierarchical' model - particularly with the metro scheme - and the local governance approach than with the public choice theory. The fact that local authorities prefer to participate jointly with other area-wide authorities (including central government units) partly explains why a fragmented response to air pollution may not be the most efficient (in terms of costs and externalities) nor coordinated way for dealing with this particular environmental issue. Nevertheless, the public good dimension that the public choice model inserts into the discussion of reorganising structures of local government contributes to a better understanding of how the functions or subfunctions of an air quality management regime (i.e. costs) can be most efficiently allocated to local units of government.

In the case of Mexico City, the extent to which delegated units believe they can solve air pollution problems is also revealed by the outcome of the survey. Again, their participation is associated with the idea of jointly working with the existing area-wide authority - the GDF - which fits with traditional assumptions. It is important to note that despite the existence of a specialised agency working on air pollution - i.e. CAM - the answers have continued to favour a body like the GDF for coordinating efforts with surrounding municipalities, monitoring systems, road traffic and the contingency plan. This in turn can be linked to the need for governmental units to manage air quality rather than non-governmental ones which operate on a cooperative or voluntary basis. While the empirical information in the case of the delegated units hardly fits with 'market' or 'network' approaches, it does suggest that there is increased interest in involving fragmented units and other actors in society in air pollution matters.

In the next two chapters, more detailed interviews with diverse key policy players further develop some of the local authority viewpoints that have been presented in this chapter. By exploring local government organisational arrangements and air quality management issues in London and Mexico City, the following chapters seek to highlight relevant issues to the three approaches outlined in Chapter 3. A comparative analysis of the outcome of Chapters 5, 6, and 7 is made in the last chapter of this book.

Notes

1 Although the current UK National Air Quality Strategy (NAQS) was not in operation during the fieldwork interviews with local authorities in London (January-May 1995), it amounted for the highest percentage (42%) of the responses.

2 For an account of London borough statutory duties regarding air quality, see Chapter 6.

6 Managing Air Quality in London

You have your own company, your own temperature control, your own music - and don't have to put up with dreadful human beings sitting alongside you.

Steven Norris
(former Conservative minister with responsibility for public transport in London) on praising the advantages of travelling by private car.

Introduction

Over the last few years, air pollution control in London has involved a mixed system of government intervention. The main actors have included an elected central government and a system of 33 unitary local authorities; other bodies, such as borough joint committees or London-wide bodies, have also participated in diverse policy formulation issues. Although the position of central government has influenced the way in which air quality has been managed in London, the limited powers of London boroughs, compared with central government, have been regarded as a limitation when attempting to deal with air quality problems (Elsom 1996: 192). While the issue of the wider participation of local authorities has been addressed as a result of the latter, the need for a strategic London-wide body specifically dealing with, for example, monitoring systems or public transport, has been put forward by such organisations as the extinct London Boroughs Association (LBA) or the Royal Commission on Environmental Pollution (RCEP), and by the Labour Party (Bell 1993: 10; Labour 1997: 34; RCEP 1994: 212-214). This chapter examines how the system of local government in London operates in relation to air quality management including the issues of monitoring systems, public transport and traffic management, and the application of an eventual contingency plan. In so doing, it highlights relevant points that reinforce diverse issues made earlier about the three main local government approaches. The analysis is based upon diverse semi-structured interviews carried out with key policy actors such as central and local authorities, London-wide bodies, and non-governmental organisations.

Managing air quality: central and local government participation

Central government has always played a significant role in air quality management; its participation has concentrated on financial, technical and personal resources that have supported and contributed to, for example, the elaboration and standardisation of air quality norms, research and educational projects, technical equipment, and the implementation of policies whenever it has been required. As shown in Table 6.1, there are many areas in which different central government agencies intervene so as to prevent and control polluted air.[1] While both the central and local government have powers in the elaboration and implementation of air pollution control policies, it is the role of central government in both areas that has predominated. Under Part 1 of the 1990 Environment Protection Act (EPA), major pollution control responsibilities have been allocated to the central agency Her Majesty's Inspectorate of Pollution (HMIP), now the Environment Agency. Under the Integrated Pollution Control (IPC) system, this agency controls the more complex polluting processes seeking to prevent or minimise pollution of any environmental medium. By contrast, local authorities control solely atmospheric releases from the less polluting processes through the Local Air Pollution Control system (LAPC). So, for example, although the number of industries in London is low compared to the rest of the country, there are still some industries that operate within the Greater London area where central government is the main author responsible for controlling pollution. By way of illustration, within the administrative boundaries of the London Borough of Brent, the local authority has no access to inspect an incinerator, a power station and the Guiness Brewery. The Pollution Inspector in this borough explained that if there is a public complaint from anyone in the borough regarding pollution from the power plant, the local authority can do nothing but to call the HMIP. In turn, the HMIP does not give feed-back about these complaints. Local authorities, thus, are confronted with the fact that, while their councillors have been elected and thus need to respond to public demands or complaints, they lack the powers to do so.

Additionally, local authority powers over prescribed processes designated for local control need to follow process guidance notes (PGs) issued by the Department of the Environment (DoE) (see Ball & Bell 1995: 292; EA 1995; EPA 1990). Although the Local Authority Unit within this government department takes the initiative to investigate what kind of guidance for local prescribed processes control should be followed in different industrial sectors, a civil servant at the Air Quality Division at the DoE explained, for example, that there is not an integrated, strategic activity focusing specifically on London coming from that central government agency. While the 1995 Environment Act (EA) and the UK's NAQS increased local authority air pollution responsibilities, the Conservative government made it clear while in power that it had 'no plans to change the principles which underpin IPC or LAPC' (DoE 1996: 68-69).

Table 6.1 DoE and DoT's main air pollution control responsibilities in London (before June 1997)

Government Department	*Main functions*
Department of the Environment (DoE)	Industrial pollution control (assessment and enforcement) Guidance notes for local authorities control over Prescribed processes Air pollution monitoring and measurement Air pollution data bank (privately contracted) Air quality standards and guidelines National air quality strategy *Others include:* Strategic planning guidance and land-use planning Liaison with London Boroughs and local authority Associations
Department of Transport (DoT)	Public transport (includes underground, buses, coaches, railways) Road schemes (building, maintenance) Traffic management (red routes, traffic control, signing) *Others include:* Transport planning Liaison between British Railways and London Transport

Source: Various

Heavy central government intervention in air quality management stems from four main factors. First, Britain has been a unitary state, and thus, local government has been formally subordinated to central government. The fact that local government has no constitutional status - i.e. can be created or abolished at will by a majority vote in the House of Commons (see Sharpe 1995c: 111-113) - has created an enormous imbalance in power and self-confidence between central and local government (Gordon 1993: 14). Although the history of British local government shows how an unchanged constitution allows variations in central or local government authority, the latter no longer has the political weight and local independence it once had before central government intervention increased in the nineteenth century (see, for example, Foster et al 1980: 21-30; John 1990: 1-5). An example of this evolving process is provided by the relatively recent abolition of the GLC in 1985, which took place in spite of opposition from the GLC and local authorities, in response to the 1983 Government White Paper 'Streamlining the Cities' (see Chapter 4).

Second, the Conservative Government's approach to local government reorganisation during the 1980s and 1990s increased central government intervention (see, for example, John 1990: 14-16). The New Right values associated with conservatism, not liberalism, put forward by the Conservatives largely influenced central-local government relationships (see also Chapter 4). Within this, central government emphasised the constitutional position of Parliamentary sovereignty over local authorities (in the case of London, the existing level of Parliamentary interest was very low):

> The Conservatives are enthusiastic about local government as long as these local authorities satisfy central criteria and aims (King 1995: 237).

Hence, as local authorities lack legislative powers - other than through making by-laws under Parliamentary delegated legislation (HMSO 1996: 4) - it is the UK government which has ultimately allocated local functions, taxation levels, standardisation of urban environmental services, and so on through national legislation. Additionally, although in 1979 the Thatcher government promised a drastic cut of quasi-governmental organisations (QUANGOS) as part of a political and economic strategy to reduce public expenditure, this has not been achieved. On the contrary, these organisations started to be seen as a convenient way for the British government to intervene at the local level without political opposition from local government:

> Quangos have provided a means of developing an administrative network at grass roots level without the uncertainty (and possible political hostility) associated with provision by elected local authorities. At the local level recent years have seen considerable central intervention via 'fringe' agencies (Greenwood & Wilson 1989: 218-219).

Third, London plays a very important role within the national political and social scenarios. Any failure in the delivery of urban environmental services in London does have a highly magnified effect on informed debate and politics. Due to its dominant position as the capital city of the UK, where a large proportion of the country's social problems and political power are concentrated, central government has to be concerned about the way London is governed (Travers et al 1991a: 54-55). Thus, it is in central government's political interests to have tight control over city management issues in London. A weak system of local government with additional financial constraints in the capital facilitates this. As already discussed, the abolition of the GLC in 1985 had the effect of weakening the upper tier. At the lower tier, although the London Boroughs gained some powers, they did not acquire greater responsibility for controlling air pollution, and in turn were left to intervene voluntarily in some areas, for instance, on monitoring air quality, a situation which lasted until 1996 when statutory duties were introduced in this area through the NAQS. Many of the borough urban environmental services are constrained and controlled by central government economic policies and local funding regimes, such as planning and public transport (see, for example, Travers et al 1991b: 5 and 1991d: 2).

Finally, it has been mentioned that central government intervention also stems from local authorities' lack of action due to their lack of interest in, or awareness about urban environmental issues (see, for example, Gordon 1993: 14). It could be argued therefore, that in the case of the London Boroughs, if their participation has been allegedly minor, this has to do not only with their lack of powers and financial constraints but also with their lack of political will. Even so, according to the results of the semi-structured interviews carried out with local authorities in diverse boroughs across Greater London, there are some London boroughs that have been actively participating in air pollution control programmes, regardless of their functional and financial limitations (see also the survey results in Chapter 5). Thus, the arguments put forward by some central government authorities that new responsibilities cannot be allocated to local authorities, due to financial and personnel shortages or political local inaction, are weakened by the initiatives taken by some boroughs such as the City of Westminster and Brent in response to air pollution problems.

Indeed, the Westminster Initiative - launched in 1989 by the City of Westminster - is a good example of how London boroughs are responding to air pollution problems. This Initiative is an environmental partnership between the Council, residents, voluntary organisations and businesses, working together to protect and improve the environment. Some of the actions in the Initiative include the 'Exhaust Watch Scheme' to combat vehicles with smoky exhausts and the adoption of the Environmental Charter which commits the Council to conserve energy. While this borough has one of the largest budgets in London, it has also secured financial assistance from businesses that sponsor different projects as partners of the Westminster Initiative (City of Westminster 1993: 63-64). Apart from this, the City of Westminster also carries out its own air quality monitoring system and continues to work on the 1994 Cleaner Air Campaign to further measures to combat air pollution from vehicles.

Another example of participation is provided by the London Borough of Brent. This borough, within the framework of Local Agenda 21, organised a series of seminars in April 1996 which depicted diverse proposals and projects from different sectors of society in order to discuss how the Council could respond to transport, environment, pollution and socio-economic issues. The initiative for these seminars came directly from the Council and allowed the participation of voluntary organisations, industrial and commercial representatives, schools and citizens. Apart from this, a real time air quality monitoring site has been installed within Brent's administrative boundaries and has been included in the DoE's national network. The initiative to set up such a monitoring site, and the origins of the funding, came from the borough. The Pollution Inspector in this borough, who is in charge of monitoring air quality in the borough, explained that if it had not been because of their initiative in going to the DoE, the DoE would not have gone to them. They contacted the DoE because the borough wants to achieve national guidelines. The DoE sends information about air quality to the borough, and the results of the monitoring at Brent are on the Internet and on Ceefax. Additionally, through its Environmental Services Department, the borough is working in partnership with Brent Traffic Consultancy, responsible for managing the traffic on Brent's busiest roads and controlling traffic and crowds on the roads leading to the busy Wembley Stadium complex. As part of its statutory powers, the Council continues to take action against smells, fumes or dust and on registering and inspecting local industries under the LAPC.

Table 6.2 Central and local authority views on the role of central and local government regarding air pollution in London

Central authority views		*Local authority views*
Intervention accepted	Central government's role: norms and regulations	Intervention accepted
Divided views on intervention	Central government's role: implementation and enforcement	Divided views on intervention
Against intervention	Local government's role: norms and regulations	Intervention accepted
Divided views on intervention	Local government's role: implementation and enforcement	Intervention accepted

Source: Author's semi-structured questionnaires

The outcome of the interviews carried out with central and local authorities in London shows some interesting results regarding central government participation. As seen in Table 6.2, according to most central and local authorities, central government intervention on the elaboration and standardisation of norms and regulations is generally accepted. Nevertheless, there is controversy both regarding whether central government should intervene in the enforcement of certain air pollution control policies or not - such as industrial emissions in London - and on the areas where this intervention should take place. Two interesting examples given in relation to the latter concern who should have the legal duty to see that air quality standards are met and who should assess air quality monitoring. According to the responsible for the preparation of the NAQS Consultation Draft at the DoE (Air Quality Management Division), ensuring that standards are met and assessing monitoring is closer to central government thinking, as opposed to actually meeting the standards or monitoring air pollutants. In practice, however, central government elaborates and assesses air quality standards - e.g. industrial emissions - as well as assessing monitoring and actually carrying out monitoring duties.

By contrast, local authorities have long requested a duty to assess ambient air quality areas and to monitor air pollutants. The interest of local authorities in getting involved in air quality management policies was confirmed in an interview carried out with the Director of Professional and Technical Services at the Chartered Institute of Environmental Health Officers (CIEHO).[2] According to this person, it is essential that local authorities are involved in air quality assurance and the collation of data from monitoring. He argues however, that a statutory duty to assess areas regarding monitoring requirements is preferable to a duty to actually monitor. The reason for this is because the former would provide for targeting financial resources better than statutory powers on monitoring. The CIEHO position not only conflicts with central government views on these issues, but also with other local authority associations, such as the Association of Metropolitan Authorities (AMA).

Before the enactment of the 1995 EA, a number of recommendations were made on granting local authorities new statutory duties for air pollution control. Such duties included assessing ambient air quality, monitoring air pollutants, banning traffic when air pollution episodes occur, and other and diverse traffic management and planning responsibilities (see Bell 1993: 8-10; FoE 1991: 56-57; RCEP 1994: 236-237). The 1995 EA has given the London Boroughs some of the powers needed to tackle air pollution in London. For instance, local authorities in London now have clear statutory powers on assessing air quality within their respective areas, as well as banning traffic temporarily along certain routes during poor air quality episodes (Ball & Bell 1995: 327; DoE 1996: 61-74; EA 1995). Through this act, and its subsequent national air quality management strategy, the UK government has at last recognised 'the need to give local authorities a greater role in managing local air quality' (Elsom 1996: 192).

Nevertheless, the new approach to air pollution control through the 1995 EA and the 1997 NAQS failed to address two significant issues during the Conservative administration (EA 1995; DoE 1996 and 1997). First, there was no clear central government commitment to financially support lower-tier authorities. In fact, the air quality strategy was criticised for failing to provide any resources to achieve the new London borough statutory duties. As Toby Harris, chairman of the Association for London Government (ALG) said:

> Mr. Gummer [Environment Secretary] is passing the buck to local government. We are being asked to make schemes to clean up vehicles self-financing by fining people. They want teams to monitor traffic, but the police have to stop the vehicles, and they want paying up front for their time. It is not clear that any of this can work. If the Government was serious they would make some pump priming available to get the system up and running (The Guardian, August 22, 1996).

Second, there was no detectable sense of political will to create a London-wide authority for air pollution control purposes. This issue however, has been addressed by the new Labour government in its recent proposals for creating an area-wide authority in London with air quality management responsibilities (see Chapter 4).

From the outcome of this section there are two relevant issues that need to be highlighted in connection with the three approaches to local government. First, the existing institutional fragmentation at local level in London has fostered central government intervention on diverse air quality management issues, that is to say, control of certain areas, and responsibilities have inescapably gone to central units as envisaged by orthodox assumptions. This situation has led to a failure in the implementation of an integrated or strategic approach to London-wide air pollution problems as well as leaving a gap at the local level regarding accountability and democracy, as conventionally understood. Second, in spite of the enhanced central government intervention, several London boroughs have been actively participating in air pollution control schemes. The initiatives taken by London boroughs reinforce the results of the survey in Chapter 5 in the sense that decentralised authorities believe that they can help solve air pollution problems within a currently fragmented situation. As seen from the examples of the City of Westminster and the London Borough of Brent, although local authorities believe they can solve problems to a certain extent, in both cases they are working in partnership with other agencies. Rather than exclusively working with an upper tier of government (as two-tier advocates would suggest), local authorities prefer working on a joint basis, including not only an area-wide authority but other sectors of society, as envisaged and recognised by the local governance approach. Under public choice assumptions, some of the borough initiatives (e.g. conserving energy, combatting vehicles with smoky exhausts, or managing traffic congestion) constitute good examples of positive externalities that are created for the benefit of other surrounding boroughs and counties. Without a coordinated action for delivering an area-wide public good, these and other boroughs who are also working within their jurisdictions may not be able to internalise the negative externalities that are produced by the lack of action or cooperation from other boroughs in the same metropolitan area. The latter becomes more difficult to solve as central government has failed to provide financial resources to the boroughs and thus equity and efficiency in the distribution of the service cannot be ensured.

Problems such as the lack of an integrated approach to London-wide pollution problems or the lack of a unit to coordinate and standardise efforts among all boroughs in London, reveals the need for an institutional channel or 'bridge' to fill the gaps created by the current organisational arrangements in London. While under traditional or network assumptions, a local unit of government could solve these conflicts - either by being directly involved or by steering and guiding, respectively - under the public choice model a bargained or cooperative arrangement among all boroughs would be sufficient. I will come back to this point at the end of the following section.

The role of the London government

Discussion of the participation of the London government regarding air pollution control needs to take two issues into account. First, the long-standing debate on the relevance of an area-wide strategic authority for London has been invigorated by current air quality management concerns. Second, as the nature of air pollution in London has changed over the last decades (i.e. road transport emissions have increased through the 1980s and 1990s) air quality management measures need to address public transport and traffic management issues (see Chapters 2 and 4). The results of the semi-structured interviews in London suggest that the way in which air pollution is currently dealt with lacks the coordination and strategic elements which are required for an effective government response. The lack of a London-wide authority has complicated the London Borough response to air pollution control when they relate to other local, area-wide or central government agencies. The latter reinforces the results of the general survey carried out with London Borough authorities (see Chapter 5). The different relations that exist between boroughs and central agencies, and between boroughs and London-wide bodies however, are not the only obstacles that may impede achieving a coordinated response to air pollution. The existing conflicts in the relations between the increased and diverse number of London-wide bodies and central authorities, as well as among these area-wide bodies, add to the difficulties in achieving coordination and integration, standardisation and representation.

Local authorities relate to central government either directly or through other bodies (for example London-wide agencies) that represent local authority interests. First, the relationships between local authorities and the central government when discussing actions for improving air quality in London vary from borough to borough. When there is a direct relationship between the borough and the central government, the level of communication basically depends either on the borough's initiative or on whether or not a friendly relationship between local and central authorities exists. For instance, the local authority in the London Borough of Brent explained that liaison between the borough and the people they know in the DoE has been very good. It was emphasised, though, that the contacts at the DoE used to work for Brent, thus facilitating such good relationships. According to the local authority, the problem is whether their views will influence ministers at the DoE or not. While the people at the DoE take their views into account, in the meetings top level ministers already have a hidden agenda regardless of local authority opinions. Although the London Borough of Brent has built a friendly relationship with the DoE, it does not have similar contact with the DoT.

In the case of other boroughs, the relationship with central government is similar or even minimal compared to Brent. By way of illustration, a senior technician at the London Borough of Hounslow, feels that there is not enough consultation with local authorities. In the case of this borough, the local authority has been left out of the decisions in relation to the widening of the A4, a busy road that leads to Heathrow airport and which crosses the borough. The borough of Hounslow, currently working on a 3-year voluntary programme to pick up information about levels of CO inside cars during driving time, had - at the time of the interview - no committees or meetings to discuss air pollution problems. Their contact with the central government has been through a number of papers or 'reports' on air pollution published and sent to them by the DoE. As with the case of the London Borough of Brent, their relationship with the DoT is extremely poor. Another example is the London Borough of Havering: the Divisional Environmental Health Officer at this borough stated that they do not have a direct relationship with the central government. For instance, when asked about their relations with the Government Office for London (GoL) (set up in April 1994 within the DoT in order to coordinate the policies on a regional basis for four government departments - Environment, Transport, Trade and Industry, and Employment) the local authority said that the role and functions of the GoL were not clear enough. Although the GoL has focused on environment issues for London - including air quality - it has had limited contact with boroughs. At the time of the second interview with the same person at the borough of Havering, two years after the GoL was set up, there was still no relationship between the borough and this central government office.

Second, another way in which local authorities relate to the central government is through London-wide bodies, such as the Association of London Government (ALG), or the South East Institute of Public Health (SEIPH). Again, the relationships between the boroughs and most London-wide agencies vary from borough to borough. Some give them support, others do not. As these area-wide agencies work on a voluntarily basis where the boroughs have no legal duty to join them, some local authorities have proved to be more sympathetic to them than others. Without a single institutional mechanism that represents the needs and interests of all 33 local authorities, diverse city-wide bodies have become the agencies that represent the views of the boroughs. Such is the case of the SEIPH, one of the most important area-wide agencies in London. This institute in association with the LBA and the ALA set up the London Air Quality Network (LAQN) in February 1993 to improve air quality monitoring in London. This London network (coordinated and initially funded by the SEIPH) assists local authorities in the operation of monitoring equipment and the evaluation of data as well as presenting in a London-wide or comprehensive way air quality information (SEIPH 1994:

2:1-2:3). The members of the LAQN, made up of four Cluster Groups - each representing different boroughs - get together to discuss air quality issues with the DoE. Nevertheless, as a voluntary body, while being independent from any government agency for reliability and accessibility purposes regarding air quality information, it lacks the financial and political strengths required for influencing the decision-making process on air pollution control issues. Overall, the major contribution of the SEIPH to air pollution control is that it has actually undertaken the task to act as an air quality coordinating unit across London.

Despite its enormous contribution to London's air quality monitoring system, the SEIPH does not seem to provide the best institutional mechanism to forward local authority views to the central government. Although the SEIPH seems to represent the views of the majority of the boroughs, not all of them participate at the same level and some have a poor relationship with it.[3] This situation is still the case even though central government has chosen the SEIPH as the institute to coordinate the air quality pilot scheme in London initiated in April 1996 under the 1995 EA and the forthcoming 1997 NAQS. Without the support of all boroughs, it is difficult to see how the SEIPH will present a complete and comprehensive picture of London to the central government - let alone a single political view of the problem and how to tackle it.

Some examples illustrate the diversity of the relationship between different London boroughs and this important London-wide agency. To begin with, the City of Westminster has experienced some problems when relating to it. According to the Client Director Built Environment at Westminster, and in charge of the Cleaner Air Campaign at this borough, their relationship with the SEIPH has been 'a difficult one'. In the past, the problem was that the SEIPH used to be run with a low budget and was too ambitious for the money being received. This limitation was confirmed by the SEIPH Environmental Health Director who emphasised that the main obstacles for carrying out its aims were the financial (they had to generate their own income) and personnel resources. While the economic and personnel situation of the SEIPH has recently improved (this body is now funded by the boroughs each giving £4 000 a year) the City of Westminster prefers working with parties that will contribute more to their own economic and political interests - especially when they relate to central government - than with a body with financial and political weaknesses.

The London Borough of Hounslow provides another example of how a borough may not properly relate to the SEIPH. Since the borough has no discussion with central government on road traffic, transport, and other issues, it is the SEIPH network that has been doing the liaison for this borough. The relationship between Hounslow and the SEIPH does not always seem to have worked properly, however. According to an officer at this borough it has been the case that the SEIPH published information about air pollution and did not sent it to them. After complaining to the SEIPH about this situation, however, the borough of Hounslow now has a better relationship with them, and information is being provided.

The officer interviewed in Havering explained that while some boroughs work with the SEIPH for monitoring purposes, others prefer to work with a private agency known as Stanger[4] - the successor of the former London Scientific Services (LSS) of the GLC. Even more, some other boroughs simply go to the private AEA at the National Environmental Technology Centre (NETCEN). For example, on the one hand, the London Borough of Brent has privately contracted with Stanger Calibration Services for the calibration of the air quality monitoring equipment and the certification of the information provided; in turn, NETCEN audits Stanger. Sometimes, NETCEN itself assists the borough if some monitoring problems arise. On the other hand, the London Borough of Barking and Dagenham works with an agency known as National Power - the electricity funding company who has close links with this borough, and they have developed an urban regeneration partnership. Although these two boroughs actually liaise with the SEIPH, they both prefer working with, and being represented by, other agencies rather than the SEIPH. Furthermore, in the case of Barking and Dagenham, National Power collects air quality data for the borough but does not give the information to the SEIPH. It seems to be that the reason why some boroughs prefer working with certain agencies rather than with others has to do with personalities. In the case of the SEIPH, it has to do precisely with the Environmental Health Director, whom some boroughs like and trust, but others do not.

Turning to the relationship between London-wide bodies and other agencies, the outcome of the interviews suggests that conflicts also exists between these area-wide bodies and central authorities, as well as among them. For instance, the Chartered Institute of Environmental Health Officers (CIEHO), a national body which represents the views of Professional Environmental Health Officers (mainly within local authorities), works with the DoE in the discussion of many environmental protection issues including the air quality management strategy. In the case of London, a London Chief Environmental Health Officers Grouping working with London Boroughs, plus the London Centre of the institute exist. The CIEHO promotes sensible coordination between local authorities and central government through professional practice. It seeks to develop good practice, coordinate strategies and influence central government policy. While their relationship with the DoE (twice a year meetings) is good, they have no contact with the DoT, past contact with whom has been bitter, and mainly regarding noise issues. When the DoE enters into conflict with the DoT, the CIEHO is usually on the former's side. At the time of interviewing, they had not developed a relationship with the GoL, either. The CIEHO's relationship with other London-wide bodies seems to be contentious regarding diverse air quality issues. In the words of its Director for Professional and Technical Services, the CIEHO used to be in conflict with local authority associations, such as the AMA, over the need for a duty to assess air quality rather than monitor and plan. They also disagree with the SEIPH because 'it only provides one model on quality control, which was built up on very limited resources', and is driven by the LBA and ALA agenda for a strategic planning agency for London. This mechanism, as explained by my contact at CIEHO, facilitates getting some equipment in place, but may divert attention from more pragmatic approaches developing elsewhere. This person believes that air quality data is being used as an argument for the creation of a strategic authority.

Additionally, the CIEHO has been closely involved in the discussion of the pilot schemes across the UK, and on the elaboration of the NAQS. As the SEIPH has been chosen by the DoE as the organisation to coordinate the review of air quality areas in London, the CIEHO liaises with them. One single representative exists who is named by the Chief Environmental Health Officers (EHO) of each borough and who receives information from the boroughs and then contacts the SEIPH and the DoE to discuss diverse air quality issues (at the time of interviewing the representative was the Chief EHO from Haringey). Nonetheless, according to the Environmental Health Officer at the London Borough of Havering, not all the boroughs go to Haringey. Again, the reasons for this are because of the different preferences and political views that exist among the boroughs.

The SEIPH and the LBA represent London-wide agencies whose relationship is good with the DoE, but who have no links with the DoT.

According to a civil servant at the Strategy and Coordination Unit within the GoL, this central government office has worked with agencies such as the former LBA or ALA. During the interview, this person mentioned that the setting up of the GoL has encouraged a lot of people in London because now 'Londoners feel there is an organisation that will be battling around for London rather than ignoring it...there is a lot expectation from customers that GoL will represent the views of London to the government'. Nevertheless, the Senior Policy Assistant at the LBA said that the latter did not have a lot of direct contact with them on air quality issues. Instead, the LBA reports were usually sent to the relevant Ministry of State, quite often getting a response from them.

While not having contact with the GoL, the SEIPH used to relate to bodies such as the LBA and ALA where, despite the political differences in each of them, consensus had been reached on issues such as the need for a more coordinated approach on air quality information and health. Although the SEIPH has emerged as a coordinating unit for improving air quality in London however, this institute avoids contact with community groups and non-governmental organisations. As its Environmental Health Director points out, the SEIPH has to be extremely selective with the groups they contact and provide information to because they lack the resources and time to respond to everybody. In seeking for financial assistance, the SEIPH contacted London First - a business community organisation working in partnership with central, local government and the voluntary sector - to see whether the private sector would get involved, but the latter has been reluctant to support what they are doing.

Contrary to what some local authorities and other London-wide agencies have experienced, London First has been working with the DoT's agency the GoL. For example, in September 1994, London First organised a campaign called 'Breathe Easy' where the DoE, the GoL, eight local authorities, and other participants (such as Sainsbury) also collaborated. This campaign, launched by Secretary of State to the Environment, J. Gummer, gave people the opportunity to have their cars' emissions tested; if they failed the test, then they were asked to rectify it. It is particularly interesting to note that, while central authorities participate in such events in order to improve their political image, financial resources play a big role in attracting central and other private agency involvement in environmental action.

Interestingly, the outcome of the interviews about the relationships between local authorities and other agencies shows that neither the boroughs, nor the London-wide bodies interviewed, relate to the surrounding counties when addressing air pollution control. So, for example, the Borough Environmental Health Officer at Kingston Upon Thames said that there is no relationship with the surrounding county of Surrey, mainly due to lack of financial resources. Likewise, in the case of Hounslow, although there is some contact with Surrey authorities, there is no discussion about air quality management issues. Although the London Borough of Enfield is surrounded by two counties, Essex and Hertfordshire, the officer at the Group Manager, Safety and Pollution Control unit stated that there is no contact with any of them regarding air pollution control. Similarly, although the local environmental officer in the London Borough of Havering stressed the need for some level of coordination between Essex and his borough, the local authorities in Havering have no discussion with this county regarding air pollution aspects. If the local authority in Havering has had some discussion with surrounding local authorities about air quality issues it is because they work together in the South East Thames Corridor. The local authorities in Essex provide the results of their air quality monitoring to National Power, which, as already seen, does not give them to the SEIPH.[5]

The outcome of the semi-structured interviews suggests that there is a general consensus among all interviewees (including central authorities) that tackling air pollution needs a London-wide strategic agency. This is because the existence of 33 different local units - without a coordinating body - with different political interests and needs and with different views on environmental matters has allowed some boroughs to be more enthusiastic or pro-active than others. The diversity that this situation brings for Greater London when coping with polluted air - i.e. each borough acting on their own - obviously leads to a non-coordinated approach to the problem, and therefore it creates a number of variables that make the issue of combating air pollution in an integrated fashion more difficult to achieve. As already seen, while some boroughs are more ahead than others in terms of actions, others seem to be better represented by the London-wide bodies they decide to join. Without a single voice to represent them all, and with the complicated liaison jungle through voluntary London-wide bodies with lots of joint work, some boroughs do get more involved than others in discussing policies and putting forward their local interests. The overall result is an uneven response from local authorities in London.

While the results of the interviews have suggested the need for an area-wide body in London, they have also shown that there is much debate about what type of body this should be. The interviewee preferences varied according to the political interests, functions and objectives of the agencies they represent, or to their own personal viewpoints. Three different kinds of preferences that relate to the type of London-wide agency that is required for coordination and strategic purposes exist (see Table 6.3). First, central authorities and some boroughs (such as Kingston Upon Thames) believe that the coordinating body should be within central government. Second, London-wide bodies (such as the LBA or London First) and other boroughs (such as Enfield and Havering) support the idea of an elected, independent, and accountable authority for London. Third, other bodies (such as the CIEHO) and some other boroughs (such as the City of Westminster) favour the idea of a London-wide body, but without showing much sympathy for an elected, strategic authority back.

Table 6.3 Interviewee personal viewpoints on the best type of London-wide agencies for air pollution control

Viewpoints by:	London-wide agency within central government	London-wide elected agency	London-wide non-elected agency
Central Government	YES	NO	NO
London-wide bodies	NO	YES	YES
London Boroughs	YES	YES	YES

Source: Author's semi-structured questionnaires

It must be noted that according to the interviewees who favour the creation of an area-wide elected authority in London, the existence of such an agency gives more advantages than having a central or non-elected agency. First, the complicated liaison web that currently exists in London may be diminished if there is only one single head who takes over diverse air pollution control responsibilities. As the Environment Health Director at SEIPH put it 'if you are dealing with a London-wide body, then you can avoid the difficulties of dealing with 33 different organisations'. The initiative would then come from the body itself and not from each borough. This interviewee explained that whether there is enthusiasm or a lack of interest from the boroughs on air quality issues, with an elected authority they would have to participate. However, as an air quality management strategy needs to include local support, the problem here is how to make all boroughs participate and decide which air pollution control functions should an elected body take away from them. With regard to this issue, the Environmental Senior Policy Assistant at the LBA, said that the creation of a strategic authority should not take too much power away from local authorities who need to work at the local level; rather, it should be 'a sort of scale-down strategic authority that has certain powers and that is able to coordinate things that would not take too much away from the boroughs'.

Second, a London-wide authority could be politically strong enough to lobby for London borough interests. The case of the recent pilot scheme for reviewing air quality management areas in London in response to the NAQS, is a good example of how borough concerns may be heard but ignored by the central government due to the lack of a strong political body to channel their demands. The Environmental Health Officer at Havering, stated that his borough will report to the SEIPH about the air quality assessment of their area, which in turn will report to the central government. This officer wants the SEIPH to back him on the issue of the lack of knowledge that exists in his borough (for example, air pollutants, health effects, monitoring) when this body reports to central government. A positive response to Havering local authority demands will depend very much on how the SEIPH presents the information to the British government and if it manages to convince central authorities of the need for more financial resources to cover this borough's demands. Nevertheless, the SEIPH is not politically strong enough to outbalance central authority views. While another level of bureaucracy (upper-tier of government) between a local authority and the central government may still make no difference, an elected London-wide body has the advantage of representing the views on air quality management issues of members who have been democratically elected. As explained by the Environmental Health Officer in Havering, having democratic local support - something that the current London-wide bodies lack - could enhance local authority interests when discussing air quality issues with central government.

Finally, another advantage mentioned by some interviewees regarding the creation of an elected authority in London is that such a body may coordinate diverse actions between boroughs and surrounding counties. While the problem here lies on how to bring all parties together into one single forum of discussion, the outcome of the interviews has shown that this task has not been achieved either by central government, or the existing London-wide bodies.

As explained at the beginning of this section, the current fragmented situation in which local authorities operate has complicated their response to air pollution when they are in contact with other governmental and non-governmental agencies. The varying relationships of London boroughs (for example, Brent, Hounslow and Havering) with central government units and other agencies or surrounding counties, that stems from the lack of a proper area-wide institutional bridge, has produced an unequal and non-coordinated response to air quality management. The latter raises the issue concerning the need to create a proper institutional arrangement in order to facilitate the relationships of all agencies involved in managing air quality.

The 'market' assumption that in a fragmented situation local units may be able to agree on cooperative arrangements and thus solve large-scale management conflicts (as outlined by the public choice theory) has not worked properly, as shown in the case of London. This is because not all the area-wide bodies that exist are voluntarily supported by London boroughs and some have developed better and closer relationships with diverse boroughs and central government units than others. On top of this, as some local authorities may financially contribute more than others, area-wide bodies are subjected to financial limitations in carrying out their functions. The costs for the boroughs that contribute more for delivering a public service are consequently higher than for those who are less willing to provide financial resources. Such is the case of the City of Westminster that not only provides funding to some area-wide institutions (e.g. SEIPH) but works with its own environmental initiatives. This supports one main concern of orthodox advocates regarding fragmentation: uneven response from local authorities exists (which in turn creates spillovers within a metropolitan area) because some boroughs are more enthusiastic or pro-active than others (see also Chapter 5). In spite of the latter, and although some of the existing area-wide bodies have failed to provide a coordinated, efficient (in terms of costs) and strategic response to air pollution, the results of the interviews suggest that there is still some outstanding debate on whether such an institutional arrangement should be an elected, non-elected, or central government agency.

It is important to note that the various agencies that exist in London for managing air quality and their relationships can fit into the local governance approach that recognises the existence of several stakeholders operating in the governing process. The dilemmas that are revealed in the latter are that many agencies and inter-local or inter-metropolitan relationships can lead to blame avoidance or scapegoating (see Chapter 3). There is a blurring of responsibilities which creates ambiguity and uncertainty among citizens about who is responsible for what. Other present dilemmas include accountability and leadership failure. This is because the current system of local government in London can only indirectly or imperfectly steer networks or partnerships as diverse agencies operate with significant degrees of autonomy. In any event, as local authorities are subject to private contractors, providing air quality management services (e.g. monitoring data) may be unequal and inefficient as the costs depend on which agency the local authority uses. Without leadership - let alone some form of command mode of governing - consensus building among many stakeholders would seem difficult to achieve and constant conflicts are likely to continue to appear.

Air quality monitoring systems

After the abolition of the GLC in 1985 and until the 1995 EA, air quality monitoring in London had been carried out either by the GLC's London Scientific Services (which closed down its four sites in 1991 due to lack of funding), the DoE, or by the boroughs on a voluntary basis. From the early 1990s, the DoE started to expand its real time monitoring equipment and the London boroughs did not only continue to monitor with manual equipment (e.g. diffusion tubes, smoke filters, SO_2 samplers) but began to buy their own continuous monitoring systems. During the fieldwork period in London, several visits were made to monitoring stations in diverse London boroughs where informal discussions took place with some of the technicians that were in charge of the operation of the system. In the cases of Brent, Camden, Havering, and Hounslow, the local authority showed and explained the operation of their own air quality monitoring stations. In addition, another visit was made to a DoE monitoring site: the Enhanced Urban Network (EUN) Phase 2 at University College London UCL which was set up in 1993 but started monitoring just after the June pollution episode of 1994. This monitoring site is now part of the Department of the Environment, Trade and the Regions' (DETR) Automatic Urban Monitoring Network (AUN).

The results that stem from these visits show that the air quality measurement system in London is not as sophisticated as it is in other cities, especially if it is compared to those operated in Mexico City, Los Angeles, New York or Tokyo (see WHO/UNEP 1992: 36). By mid-1994, there were approximately 15 local authority continuous monitoring sites measuring diverse pollutants quality,[6] and around 8 - some integrated from local authority sites e.g. at Bexley and Eltham - that corresponded to the DoE's network. Over the last few years, the number of real time stations acquired by the London boroughs have increased and some have become part of the DoE's air quality network. The initiative and funding for acquiring these monitoring stations has come from the boroughs themselves. For example, a real time monitoring site was set up in April 1996 in the London Borough of Brent and it has already been affiliated to the DoE's network system. By the time the Consultation Draft of the UK NAQS was issued - August 1996 - local authority automatic sites in London at Bexley, Brent, Eltham, Haringey (2 sites), North Kensington, Sutton (2 sites), Tower Hamlets and Wandsworth had been integrated into the national network. By mid-June 1997, 15 of the air quality monitoring sites of the London Air Quality Network had been integrated into the DETR's Automatic Urban Network. At the end of 1997, the monitoring network in London was made up of around 47 local authority sites and 7 DETR sites (see ALG 1997: 6; Bell 1993: 28; DoE 1996: 81; EA 1995; NETCEN 1995: 21-26; SEIPH 1994: 2/1-2/8).

During recent years, there have been three major criticisms of London's air quality monitoring system management. To begin with, criticisms have focused on the lack of a coordinating agency both to link the national air quality monitoring network with the local monitoring one, and to provide standardisation in air quality databases. First, the importance of linking the national networks with the local ones is necessary in order to obtain valuable additional information making the best use of both national and local resources (see, for example, Bell 1993: 26-33; LRC 1993: 134-137; QUARG 1993: 3-27). It has been argued however, that the British government has treated local authority monitoring 'as an optional add-on rather than the basis for building a coherent national network' (Brown 1993: 3). While central government has recognised the significance of a coordinated approach to air quality monitoring and of local monitoring for a harmonised quality assured network (see Brown 1993: 3; DoE 1996: 81), it has not allocated additional resources to the London Boroughs for acquiring new equipment. The integration of local authority monitoring may be a sensible and cost effective means of expanding the national network, but it could be perceived as 'little more than an on-the-cheap and piecemeal response to evident public concern about UK air quality monitoring' (FoE 1994: 3). If, at the time of writing, more local than central real time monitoring sites exist in Greater London, it has more to do with the initiative and local funding of the boroughs than the financial support from central government or from any of the London-wide bodies to shoulder local authority expenses. On top of this, most of the time, the boroughs are the ones that need to request affiliation to the DETR's network; such was the case of the London Borough of Brent.

Second, since the existence of the GLC, there has been a need to establish a comprehensive air pollution database for London. While such a database was set up by the GLC's London Scientific Services during the early 1980s, the GLC's Pollution Monitoring Group liaised with all organisations involved in air pollution monitoring for data collection and standardisation purposes (GLC 1983: 35). As the GLC was abolished, air quality databases - before the creation of the LAQN in 1993 - were not standardised or comprehensive. Without an area-wide coordinating agency, most monitoring stations were working in isolation and data was not systematically pooled. Despite the existence of a monitoring scheme for a range of pollutants coordinated by Rendel Science and Environment (later known as TBV Science), not all the boroughs participated and thus coverage was incomplete (see Bell 1993: 28-29; SEIPH 1994: 2/1). The underlying reason for having a coordinating agency is that monitoring on a London-wide basis provides a comprehensive and detailed analysis of air quality giving 'a strategic overview of air quality across London' as well as 'data for individual areas' (Bell 1993: 26). A comprehensive databank allows authorities to assess whether pollution control measures are stringent enough for the protection of human health, flora and fauna and the environment. In order to develop such a comprehensive approach, it is important to ensure common monitoring standards. This can be achieved by using the same methods and frequency of measurements, location of sites, types of pollutants, analysis of results, quality of equipment and calibration, and so on (see DoE 1996: 80-83; GLC 1983: 35; SEIPH 1994: 1/5). Thus, a comprehensive and standardised monitoring system contributes to knowledge about pollution sources, human health effects, the prediction of outbreaks of pollution episodes and dispersion of pollutants, and overall, to more reliable and accurate information. Although the SEIPH - through the LAQN - has already done a comprehensive audit of all local authority monitoring sites, and created a database of sites, types of monitoring systems, pollutants measured, etc. (several reports have already been published), not all the boroughs, as already seen, have contributed in the same way.

Another criticism of London's air quality monitoring system is with regard to local authority statutory powers. The main concern for most local officers before the 1995 EA was enacted was the fact that London local authorities had no statutory duties to monitor or enough financial resources to do so. The officer interviewed in Enfield explained that while air quality control was not a statutory duty, and thus the borough did not have any obligation to provide information to the public, they still had to do it because of citizen complaints or because it was requested. As they had no legal duty to monitor, this was not a priority and consequently finance was given to other areas; as Bell (1993: 28) pointed out, 'cost is a major obstacle to local authority monitoring. It is a non-statutory function which makes it particularly vulnerable to cuts in spending'. According to an officer in the London Borough of Havering, one of the huge advantages that the former GLC had in this respect was that they were interested in monitoring and had their own financial resources to buy equipment and set up databanks.

Now, after the 1995 EA, the London boroughs have a duty to provide an air quality assessment of their areas which implies a duty to monitor. As already seen however, local authorities have not been allocated financial or personnel resources for this yet. In a second interview with the Environmental Health Officer at the London Borough of Havering, I was told that the borough is trying to start monitoring PM_{10}, a pollutant of much concern for them. The new equipment, only to continuously monitor PM_{10}, costs £27 000. Due to financial constraints, as this officer explained, the boroughs cannot cover all pollutants, and thus, end up choosing the ones they believe are the most important to monitor. Thus, the monitoring results that the boroughs send to SEIPH or NETCEN, are based on non-comprehensive borough criteria for the selection of the pollutant they monitor. According to an officer at Brent, as boroughs now have a duty to monitor, they have to put more resources into it, which seems positive, but without adequate central government financial and personnel assistance (she is the only one carrying out monitoring in the borough), it has become 'too much' work for them.

Finally, the fourth major criticism of London's air quality system is related to the DoE's monitoring network. The DoE's limited number of sites and their particular locations - mainly background as opposed to kerbside and suburban - have not been sufficient to provide the comprehensive data required for the Greater London area. It has been argued that background sites may regularly miss higher pollution levels in areas of heavy traffic and risks misleading citizens about the real levels of pollution (Bell 1993: 26-33; Dobson 1995?: 4; FoE 1994: 3; SEIPH 1994: 1/3).

These criticisms have certainly highlighted the need for having an air pollution-wide agency where London boroughs can participate and contribute to the air quality monitoring system. Again, the question is whether this agency should be within central government, within an elected local authority, or simply be a non-elected one. In an interesting interview in the London Borough of Enfield, an officer pointed out that if 'we have subscribed and agreed to work with SEIPH...[it is]...because it is the only agency that wants to analyse and evaluate the problem London-wide, regardless of their motifs. So, if they are offering to do that work, then we join them'. It might well be the case that other boroughs have done the same because of a lack of any other coordinating unit. As importance and awareness of a comprehensive strategy for air quality monitoring has grown among local authorities, the creation of a London-wide authority ensuring standardisation of information and financial and personnel support would encourage London boroughs to join in and work together.

The case for monitoring air quality in London as presented here raises the question of whether providing public goods under a fragmented situation (i.e. market approach) may fail to exploit large economies of scale. Because London boroughs acquire monitoring equipment depending on their financial situation and on their political interests or priorities, the information provided is incomplete and not systematically pooled; that is to say, it fails to give a metropolitan-wide view of the problem. This situation reinforces the results obtained from the survey in Chapter 5, where even with statutory duties to monitor, money can be sent to other areas (see example of Enfield), producing an unequal service provision, which in turn may create externalities across the metropolitan complex. At present, boroughs are monitoring those pollutants they believe are most dangerous, and with diverse types of equipment, which provides inefficient and unequal air quality information for the whole metropolitan area.

Transportation and traffic management

As already explained, road traffic is the main source of air pollution in London, accounting for virtually all the emissions of energy-related hydrocarbons, CO, black smoke and the majority of NO emissions. It is also the major source of lead and other particulates, as well as an important contributor to SO_2. Although O_3 is a secondary pollutant, the pollutants that synergistically contribute to its formation - NO_x and hydrocarbons - mainly come from road traffic as well (see Chapter 2).

Whereas road vehicles are considered to be the major source of pollution in London, not all types of vehicles pollute in the same quantity. As seen in Table 6.4, private cars are top on the list of polluters by energy use in London; this table gives an estimate of how much different types of road vehicles may contribute to air pollution.

Table 6.4 Energy use in London by road vehicles

Road vehicles	*Energy use (%)*
Automobiles (private cars)	56.4
Light goods vehicles	17.8
Medium goods vehicles	11.2
Heavy goods vehicles	8.4
Buses	4.1
Taxis	1.4
Motorcycles	0.7

Source: LRC (1993)

While private vehicles have become the main source of most pollutants of concern in today's London atmosphere, they amount for less than half of the total journeys in London by mode of transportation (see Table 6.5).

The outcome of the semi-structured interviews carried out in London shows that the main air pollution source of concern for local authorities is road vehicles (see Chapter 5). This concern arises mainly from the number of public complaints that the London boroughs receive on, for example, smoky cars. For instance, the person interviewed at Enfield explained that during the period from 1994 to the beginning of 1995 (aprox. 18 months) this borough received over 100 calls alone complaining about smoky cars. Similarly, while in Havering most complaints concern vehicle emissions because of the M25, in the borough of Kingston upon Thames my contact pointed out that air pollution accounts for 50 to 100 complaints a year, basically from road vehicle emissions.

Table 6.5 Journeys in London by mode of transport

Mode of transport	*Distribution of journeys (%)*
Cars	37.0
Buses	12.0
Underground	5.5
Railways	3.0
Other	4.0
Walking	38.5

Source: LRC (1993)

In the case of the borough of Enfield however, an officer at the Group Manager, Safety and Pollution Control stated that there have also been air quality complaints regarding high levels of pollution that do not specifically originate in the borough itself. This borough tends to have problems with polluted air that is brought by the winds from the centre of London into its geographical boundaries. The centre of London is where most concentrations of pollutants are higher than in other areas - e.g. SO_2, black smoke, CO, NO_x, VOCs - presumably because this area has a major traffic flow (LRC 1993: 129-132). In other boroughs, such as Havering, the south westerly winds during the summer time take all the pollution that originated from the M25 away from the borough. As a result of this, Havering authorities have found that levels of ozone are higher outside the borough, approximately 40 miles away in the countryside

Given their importance for an adequate air quality management strategy, public transport and traffic management issues have recently become the most significant reasons for reorganising local government structures in London. This is not to say that in the past these issues were not taken into account when discussing London's government organisation. By way of illustration, since the Herbert Commission was established in the late 1950s, it was clearly stated that traffic management issues should belong to the upper-tier of government in order to avoid central government assuming dictatorial powers with regard to this (for details, see RCLG 1960: 203). Later on in 1984 when the GLC responded to the Government's White Paper 'Streamlining the Cities' (DoE 1983), traffic management and planning were included as strong arguments for the need of a London-wide authority for air pollution control. In this response the importance of an area-wide authority for London-wide prevention activities was stressed:

> The Council's role is to maintain an overview of London's air quality, and as strategic planning authority it must watch the existing levels of air pollution. This role is particularly important because of London's size and the scale and distribution of polluting sources, many of which are not amenable to local borough council control. Examples of this include traffic emissions... As highway authority, the Council has always paid great attention to minimising the effects of air pollution and other impact of new road schemes and traffic management measures (GLC 1984: (31) 2).

Over the last few years, there have also been continuous calls by different organisations to create an area-wide body for public transport management in order to have strategic and coordinated policies across London (Sharpe 1995c: 124-125). In March 1990, for instance, the Chartered Institute of Transport proposed a transport authority with powers over the financing and control of London Transport (LT), British Railways (BR), suburban services, and trunk and local roads (Ridley and Travers 1991: 175). Another example is the Royal Commission on Environmental Pollution which emphasised in its Eighteenth Report the need for a body to take a strategic overview of transport problems and solutions and to have them implemented by the transport providers, whether public or private (RCEP 1994: 213). Although the Conservative government did actually recognise the need for a coordinated and integrated transport policy, it ruled out the possibility of creating a strategic transport authority stating it to be both 'undesirable in principle and unworkable in practice' (DoT 1996: 126). While the British government at that time showed some concern about the impacts of transport on air quality in London, it failed to include the creation of an area-wide body for such aims in its integrated transport strategy stressing that all transport responsibilities should lie with the government itself, LT, and the London Boroughs (DoE 1994: 49-54 and 1995: 23-28; DoT 1996:

1-5; 126-127). More recently, the Labour Government's proposals for reorganising the system of local government in London have included the creation of a strategic authority - a Greater London Authority - which will take over diverse transport and traffic management responsibilities (DETR 1997b and 1998).

Table 6.6 Interviewee consensus and disagreement on public transport and traffic management issues

Consensus	
Public transport and traffic management policies	An integrated strategy Improvement of public transport More cycling routes and better pedestrianisation Priority to public transport rather than car road schemes Reduce traffic congestion No Draconian measures Main issue of concern: motor vehicle emissions
Local government system	Coordination among boroughs is necessary
Education and social policies	Change people's attitudes: shift from cars to public transport
Disagreement	
Creation of public transport and traffic management unit (diverse agencies)	Central government agency Elected upper-tier of government Specialised non-elected agency
Allocation of functions (devolution of power to local authorities)	Air quality monitoring (obtained through EA 1995) Assessment of air quality areas (obtained through EA 1995) Testing vehicle emissions (obtained through EA 1995) Stopping (one-by-one) vehicles Banning vehicles (obtained through EA 1995 - only ban temporarily, and along certain routes during poor air quality) Major roads and IPC industrial emissions control
Use of private cars	Lack of public transport, safety reasons, consumerism, conmfort, private space freedom, selfishness and social status

Source: Author's semi-structured questionnaires

Although these two areas - transport systems and traffic management issues - are in many ways inter-linked for the controlling of air pollution, a variety of views on developing policy strategies exist. On the one hand, the results of the semi-structured questionnaires show that most of the interviewees coincide in the need for a more coordinated system among London boroughs for transport and traffic management policies. On the other hand, the debatable issues focus on the type of coordinating agency, on the issue of devolution of power to local authorities, and on the reasons why people are increasingly making use of vehicles, and thus on the policies that should be developed to encourage them to change to public transport (see Table 6.6). From an institutional point of view the controversial issue lies primarily, once again, on whether a coordinated and strategic approach should come from an area-wide central agency, a local elected body, or a non-elected one.

With the new powers they have recently acquired with regard to traffic management issues, London local authorities still face two main problems. First, the central government has not allocated financial resources to them; second, there is no coordinating agency for when local authorities decide to implement and enforce these local powers. With the existing poor relationship between the boroughs and the DoT it is difficult to envisage that a coordinated and integrated approach to air pollution when implementing traffic management policies can take place.

The Divisional Environmental Health Officer at Havering explained some of the problems faced after the enactment of the 1995 EA. While for the first time the NAQS allow Havering authorities to review and assess major roads - in particular the M25 that crosses the borough - through a duty regarding assessing air quality management areas, there is no way they could possibly control this motorway without a coordinating body if high levels of air pollution occur. If they were to implement their duties separately, for example, temporarily banning cars without getting in touch with all the other London boroughs and surrounding authorities the M25 crosses, traffic chaos would be created.

The low level of communication that exists between the DoT, the DoE and the boroughs certainly leads to a non-coordinated response to control traffic flow. Although central government has insisted that it consults London local authorities, the DoT does not seem to be included as part of this. For instance, during the second interview with the environmental health officer at Havering in November 1996, the DoT had still not contacted the London Borough of Havering to develop the transport plans and programmes (TPPs) in order to establish 'a common approach to reducing pollution and improving the environment across the capital' (DoE 1995: 25).

The fact that it is not appropriate or convenient to close roads without consulting other authorities that might be affected by such a decision was addressed by the local authorities in the boroughs of Brent, Camden, City of Westminster, Enfield and Havering. The underlying reason for this is the same: closing one road within the boundaries of a borough would divert traffic to other adjacent boroughs, creating traffic jams and thus leading to increasing levels of air pollution.

Public transport and traffic management issues constitute examples that are relevant to the public choice approach because of the externalities that can be created by these subfunctions with broad catchment areas. The case of the London Borough of Enfield, for example, where air quality complaints are about levels of pollution that have not originated within the borough's boundaries, shows the existence of negative externalities produced by other boroughs within the same metropolitan area. Under 'market' assumptions the issue of concern, in the example of Enfield, is whether the boroughs where pollution originates (i.e. road transport emissions in central London) can actually internalise the effects that are caused for Enfield. Another example of spillovers under a fragmented system are the concerns raised by the boroughs of Brent, Camden, City of Westminster, Enfield and Havering on the issue of closing roads (as well as banning or stopping cars when air pollution episodes occur) without consulting other local authorities that might be affected by that measure. The externalities created in these cases may be associated with increasing levels of air pollution as closing roads in one borough diverts traffic to adjacent boroughs, eventually creating traffic jams.

While the interviewees emphasised the need for a strategic and coordinated approach to public transport and traffic management, they also raised concerns regarding the need to make citizens participate in air quality management issues (e.g. to encourage a modal shift from private cars to public transport). Under local governance assumptions such participation could be achieved through diverse organisations (such as community groups, non-governmental organisations or public-private networks) which could encourage drivers to make less use of their cars. As yet there is limited exploitation of these opportunities.

Emergency plan

One of the first alert system schemes in the world was devised for London after the 1952 smog pollution episode ocurred; this scheme, an emergency hospital alert plan was, however, never brought into operation (Elsom 1996: 91). At the time of writing, the British government had not yet designed an alert system for London to address warnings and emergency situations when air pollution levels rise and breach the DoE's air quality limit threshold guidelines.[7] The UK warnings on high levels of air pollution in London do not include any legally-binding measures; it only requests voluntary cooperation from local authorities, drivers, etc. (see Elsom 1996: 99; FoE 1994; Weir 1993: 13). While local authorities in London now have statutory powers to restrict the use of roads during air pollution episodes (e.g. ban cars), they first need to go through a consultation process for this and prepare contingency plans (DoE 1995: 19 and 1996: 54; EA 1995).

During the fieldwork carried out in London, some interviewees explained that under the current air quality situation an emergency plan is not required. For example, the Director of Environmental Health at the SEIPH stated during an interview that the levels of pollution in London are not cause for an emergency planning situation: until November 1994 the last pollution episode of concern had occurred in December 1991 when NO_x increased and breached the air quality norms. On the other hand, the senior policy assistant at the LBA said that at an emergency level action needs to be taken more within a long term framework than a short one. According to this interviewee, there should at least be some kind of emergency banding to best inform people about the situation and perhaps encourage them not to drive into work. The central government authorities interviewed also showed more concern regarding the existence of the right air quality bands than regarding an immediate response to air pollution through an emergency plan for London.

An emergency situation is certainly subjected to the type of air quality band system that exists so as to measure whether the limits are being breached or not. The way in which air pollution concentration is banded in London raises some questions on whether there have not, in effect, been pollution episodes. For example, if an air quality band is 'too tolerant' then it is less likely that pollution episodes, i.e. an emergency situation, may occur. As the results of the general survey showed (see Chapter 5), the DoE's air quality band has precisely been criticised for being too tolerant. It has also been argued that because the monitoring network is unable to provide appropriate local information, and the air quality information broadcast is often unclear about predicting an episode, it is not possible to create a preventative plan (Weir 1993: 12), let alone have reliable information on the real levels of pollution. As already seen in this and other chapters, the number and location of the air quality monitoring sites also contributes to the latter.

If local authorities were to elaborate contingency plans on their own, these plans would lack the integrated and area-wide perspective that is required for London. Whatever the type of measures contained in such emergency plans, a head is required in order to have an overview of pollution episodes in London, as well as to coordinate all boroughs in the implementation of the 33 diverse plans. This coordination is necessary, especially if local authorities decide to control cars and close some roads. In addition, as much as the media is extremely important in the quick dissemination of the outbreak of a pollution episode, a coordinating body would certainly contribute to pass all the information to local authorities, schools, police, hospitals, and so on.

In order to know the type of response London boroughs have when an air pollution episode occurs, the semi-structured questionnaires sought to depict local authority experiences regarding the 1994 summer pollution episode in London.[8] The answers of five different geographically located boroughs at the centre, north, east, west and south of Greater London - Camden, Enfield, Havering, Hounslow, Kingston upon Thames - were quite similar: they did nothing about it. The reasons that local authorities gave with regard to this are as follows:

- Camden. They did nothing because local authorities do not have the capacity to deal with that; they first had to get in touch with health authorities to see if there was an increase of asthma or lung illness reports. After the pollution occurred they issued a press statement.

- Enfield. The reason was that, as they are not able to generate their own data, they did not know that the pollution episode would occur. They only received a fax with information about air quality from SEIPH.

- Havering. It was explained that the authorities 'did not know what was occurring'. The borough first knew about the episode from medical people, who rang them and told them about an unusual number of people going to hospital: all of a sudden hospitals were full of asthmatics. The problem was that there was no information circulating; no one told them about the situation and as a result, the local authority did nothing.

• Hounslow. Although they now work with the SEIPH, the borough did not at that time have the mechanisms to publish data, and therefore to give the information and alert the public. Emphasis was put on the need to establish the way in which messages reach people. In addition, the local authority complained that the media seems only to report when poor air quality conditions emerge.

• Kingston upon Thames. The answer was 'because we didn't know that an episode was happening'. The reason for the latter is because they did not have continuous monitoring systems to quickly identify high levels of pollution. Not even hospitals contacted them when the pollution episode took place.

As clearly seen with these examples, Greater London lacks a body to coordinate and send information to local authorities, the media, and the public - especially to vulnerable groups such as asthmatics, the elderly, children - of an eventual air pollution episode occurrence. According to the local authority in the London Borough of Hounslow, in order to give an alert about high levels of pollution it is important to follow a 'cascade system', i.e. a system where there is a head No. 1 who rings No. 2 who has different channels; then each of these ring their next channels, and these the rest, and so on. Information may be sent through faxes, via INTERNET, via e-mail, etc. After all, according to this local authority, what people want to know is when air quality is poor and not when it is good.

Although it has been difficult for the London boroughs to participate in emergency situations, some of them have already taken the initiative to do something about this particular issue. By way of illustration, the City of Westminster has included in its 'Westminster Initiative', a system to predict weather conditions and see whether poor air quality will arise or not. When poor air quality is predicted, they send faxes to schools, GPs, and chemists, in order to warn them in advance. This experiment only started in July 1995 so the effectiveness of the results cannot be thoroughly assessed yet. Likewise, the London Borough of Brent has already set up a system in which, when a pollution episode occurs, they send a fax to schools and hospitals warning them about the situation. In this Borough, when the warning has been given, schools interrupt and postpone children's exercise or physical activities.

The relevant point in this section, in connection with the three approaches outlined in Chapter 3, is that the implementation of an emergency plan when pollution episodes occur requires much coordination and consensus among all participants involved. Whatever the organisational arrangement (hierarchical, market, or network) the unit in charge needs to serve as a forum for deliberation on the type of measures to be implemented and as a centre to coordinate and disseminate information on the outbreak of a pollution episode. What is clear from the experiences of the London boroughs during the 1994 summer pollution was the lack of a body to inform them about its occurrence. Under market assumptions the issue of concern is whether all boroughs (and other participants) would actually reach consensus and cooperate equally in the implementation of the emergency plan. Under governance assumptions the question is one of the blurring of responsibilities; that is to say, who would be responsible if the alert plan was not to be launched on time or if the agreed measures during the decision-making process were not to operate properly.

Conclusion

The outcome of the semi-structured interviews carried out in London suggest that there is wide consensus among all the interviewees on the need for a coordinating and strategic body for managing air quality issues, particularly for monitoring systems, transport and traffic management, and the eventual creation and implementation of a contingency plan. However, different views exist on whether this body should be a central government unit, a local elected body, or a non-elected one. It is clear however, that air quality management concerns have invigorated the discussion for reorganising the local government system in London and that such reorganisation must include the creation of a kind of strategic area-wide authority. The lack of a city-wide authority in London has implied heavy intervention by central government units and the emergence of a number of area-wide bodies which have participated in diverse air quality management issues. On the one hand, while central government participation seems necessary (especially regarding economic and technical aspects), it has failed to adopt an integrated approach to air pollution and to create a central agency to coordinate London boroughs and other area-wide bodies in dealing with air pollution. On the other hand, London-wide agencies have not been the institutional forum (where all parties can participate) for discussing air quality management issues, nor have they become the politically legitimate bodies to put forward local authority concerns to central government.

From the outcome of the semi-structured interviews it is also possible to conclude that managing air quality in London under a fragmented system of government has not properly achieved, as traditional advocates would argue, the greatest efficiency in production nor the greatest equity in distribution. The evaluation of the empirical information in public choice assumptions suggests the substantial amount of externalities that are likely to appear in a metropolitan complex with a jurisdictional fragmentation of local units with regard to goods or services with broad catchment areas. While the participation of diverse sectors of society (as defined by the local governance approach) is frequently considered desirable, the various agencies that operate and their consequent relationships for managing air quality have resulted in complex governmental response to air pollution, which is problematic from coordination, efficiency and standardisation viewpoints.

The next chapter provides an insight of how a different system of local government operates in relation to air quality management. By following the same methodological process carried out for London, the case of Mexico City reviews many of the issues that have been addressed in this chapter. The final results of both case-studies is presented in the last part of this book.

Notes

1 Different levels of coordination among the Government Departments regarding air pollution control exist. A detailed analysis and explanation of the diverse ways in which they liaise within each other is not within the scope of this research.

2 The interests and position of CIEHO are explained later in the chapter.

3 During the interview with the Environment Director at the SEIPH, I was told that when the LAQN was set up, the network was supported by the LBA and the ALA. Nonetheless, not all boroughs supported the SEIPH in carrying out its tasks; according to this person, 31 boroughs gave their support, one answered that it was not sure, and one simply said no.

4 When the GLC disappeared, the agency in charge of collecting air quality information was Rendell Science and Environment, who then became TBV Science. The latter became a private company called Stanger.

5 It is important to note that the regional body South East Regional Planning Conference (SERPLAN) was not mentioned at all in any of the interviews.

6 It is important to note that most of these real time monitoring sites principally measure only four pollutants: NO_2, O_3, SO_2, and to a lesser extent, CO. Only two sites measure PM_{10} and hydrocarbons, and only one measures TSP.

7 This section does not consider emergency plans in the case of industrial accidents. The focus of debate is on pollution outbreaks stemmed from road vehicles. For some information on unexpected events of hazardous pollutants see the Planning (Hazardous Substances) Act (Planning 1990).

8 There is some debate on whether the 1994 'pollution' episode that took place during the summer of that year - July 15 - was really a pollution episode or a simple environmental phenomenon, that is to say, a thunderstorm. However, the outcome of local authority views and data collected during the fieldwork in London, suggests that it was, indeed, a summer pollution episode (see also, Elsom 1996: 99).

7 Managing Air Quality in Mexico City

And when we saw so many cities and villages built in the water and other great towns on dry land and that straight and level causeway going towards [Tenochtitlan], we were amazed and said it was like the enchantments they tell of in the legend of Amadis, on account of the great towers and [temples] and buildings rising from the water, and all built of masonry, and some of our soldiers asked whether the things that we saw were not a dream...

Bernal Díaz del Castillo - 1519?
(soldier and chronicler of the discovery and conquest of Mexico City)

Introduction

Over the last decade, government response to controlling air pollution in the MZMC has involved a mixed system of tiers of government whereby the main policy formulation and implementation actors have been an elected central government, an elected federated state - the surrounding State of Mexico (EdoMex) - and a non-elected Mexico City government agency: the DDF. As in the case of London, the participation of central government in managing air quality in Mexico City has concentrated on financial, technical and personnel resources. Such intervention has supported and contributed to, for example, the elaboration and standardisation of air quality norms, research and educational projects, technical equipment, and the implementation of policies whenever required. Although by the late 1980s the Federal Constitution established a decentralised system for environmental management purposes, incorporating the participation of local and municipal authorities regarding air pollution control, heavy central government intervention has continued. While the participation of Mexico City's local authorities in controlling air pollution has increased, air quality management concerns have not been included in the discussion for reorganising the structures of local government in Mexico City. Based upon diverse semi-structured interviews carried out with various policy actors (e.g. central and local authorities, NGOs), this chapter examines how the system of local government in Mexico City operates in relation to air quality management. As in the previous chapter, it also seeks to highlight relevant points in connection with the three approaches outlined in Chapter 3 and to reinforce some of the results analysed in Chapter 5. The analysis of the empirical material includes the issues of monitoring systems, transportation and traffic management, and the application of the existing contingency plan.

Managing air quality: central and local government participation

From the 1930s up to 1988, air pollution matters had been exclusively a function of the federal government. While the 1987 Constitutional amendment of article 73 XXIX-G and the 1988 environmental law - the General Law for Ecological Equilibrium and Environmental Protection (LGEEPA) *Ley General del Equilibrio Ecológico y Protección al Ambiente* - allowed the government of Mexico City to get legally involved in the prevention and control of air pollution, central government has continued to play a significant role in air quality management (see Campos Ruíz et al 1993: 60; Collins & Scott 1993: 119; Ward 1990: 63-91). The heavily centralised intervention of the federal government has however been regarded as one of the main factors which has impeded an adequate response to air pollution in Mexico City:

> Although most of the environmental problems in the Basin of Mexico have reached critical proportions in the late twentieth century, industrial development is not solely to blame. Urban primacy and political centralism have been a tradition in Mexican society since the Aztec empire. The Basin of Mexico...has used its preeminent administrative and political position to obtain advantages over other areas of the nation (Ezcurra 1990: 586).

As shown in Table 7.1, there are many areas in which different federal agencies have intervened so as to prevent and control air pollution in Mexico City. For instance, the central government agency for environmental issues - SEMARNAP - has powers for regulating all fixed sources which produce pollution in the DF (with the exception of service establishments), as well as regarding the organisation and management of national parks and ecological reserves in Mexico City (see LGEEPA 1988). While both federal and local authorities legislate on the issue of air quality management, it is central government's role in this area that has predominated. By way of illustration, whereas the 1988 LGEEPA says that the DDF has powers on any type of moving sources within its territory (and on fixed sources which are service establishments) the federal agency SEMARNAP is in charge of issuing the norms for the maximum levels of emissions of such moving sources and for the norms on the quality of fuels, as well as on the elaboration of the contingency plan. The local environmental act for the Federal District - the Federal District Environmental Law (LADF) *Ley Ambiental del Distrito Federal* - enacted by the ALDF in 1996, regulates all moving sources within Mexico City following federal guidelines to this end (see LADF 1996).

Table 7.1 Functions of different federal bodies on air pollution matters in the MZMC

Government Department	*Main functions*
Secretariat for the Environment (SEMARNAP)	Industrial inspection and control (fixed sources) Elaboration of environmental norms
Secretariat of Health (Ssa)	Epidemiological supervision and health impact evaluation and citizen orientation
Secretariat of Education (SEP)	School regulation during contingencies Environmental education programmes
Secretariat of Communications and Transport (SCT)	Transport emission control (other than MZMC vehicles)
Secretariat of Commerce and Industry (SECOFI)	Change-over to gas programme Norms issue for gas use and distribution
Secretariat of Energy (SE)	Energy saving programme
Secretariat of Finance (SHCP)	External finance and fiscal measures Price policies on fuels
State-owned oil industry (PEMEX)	Fuel quality improvement Research project: Los Alamos/IMP
State-owned electricity industry (CFE)	Emission control in thermoelectric plants

Source: CMPCCAVM (1994a); LGEEPA (1988)

While the legal system determines the participation of central and local authorities, due to the imprecise distribution of functions among the federal and local government, the current environmental legislation has created enormous confusion about the elaboration, implementation and enforcement of air pollution regulations in the MZMC. While the Federal Constitution clearly states in article 48 that any issue related to the atmosphere is a federal function, it also allows local authorities to intervene on diverse air quality issues. The coordination agreements that are set out in the environmental legislation allow local authorities to only get involved at the administrative or enforcement level and regarding certain areas. Thus, formally speaking, the local legislature in Mexico City - the ALDF - does not have constitutional powers to enact legislation on air pollution issues except those that are related to the implementation and enforcement of air quality norms, with the exception of fixed sources (Brañes 1994: 123-126; Cancino Aguilar 1994: 115-136; Constitution 1997; EGDF 1994; González Márquez & Cancino Aguilar 1994: 32-42 and 63-65; LGEEPA 1988). While not clearly established, it is therefore important to distinguish between the regulatory and administrative functions (implementation and enforcement) in relation to air pollution aspects. Although Mexican legislation gives exclusive powers to the federal government regarding air quality norms, in practice federal authorities are continuously working on their elaboration and discussion, in coordination with Mexico City local authorities, and the surrounding federated state EdoMex through informal meetings.

The views of most federal and local authorities also stress the issues of confusion and overlapping of functions that derive from the current Mexican environmental legislation on air pollution. For example, the Director for Environmental Policies and Norms at the DDF, explained that in the case of fixed sources in the MZMC, different authorities intervene for the application of different environmental regulations for the same industry. If it is about air emissions, the federal authorities intervene, but if it is about water pollution, then the local ones are in charge. What is more, the generation of dangerous waste is a federal issue, but the generation of non-dangerous waste is local - and municipal if the case involves drainage in one of the municipalities of the MZMC. All this creates confusion for industry in the application of environmental regulations, resulting in two or three simultaneous administrative procedures for licences, sanctions, payment of fines, and so on. The economic costs and delays that stem from this system are substantial.

According to the Director for Environmental Policies and Norms at the DDF, the federal government should specify the norms, and the local authorities should implement and enforce them. Only when the process of implementation demands federal assistance should both local and central authorities participate.

Table 7.2 Central and local authority views on the role of central and local government regarding air pollution in Mexico City

Central authority views		*Local authority views*
Intervention accepted	Central government's role: norms and regulations	Intervention accepted
Divided views on intervention	Central government's role: implementation and enforcement	Against intervention
Intervention accepted	Local government's role: norms and regulations	Intervention accepted
Intervention accepted	Local government's role: implementation and enforcement	Intervention accepted

Source: Author's semi-structured questionnaires

As shown in Table 7.2, the argument that the federal government should intervene in the elaboration of the norms - such as environmental standards, main environmental policy, regulations on air pollution related issues such as health and vehicle emission norms, and so on - is shared by both federal and local authorities. However, according to the interviewee answers, this does not imply that the federal government should have exclusive powers over it. On the contrary, federal and local authorities views are that the central government should work together with the DDF and the EdoMex authorities on the elaboration and discussion of policies and norms for the MZMC; therefore, some legal changes are required at the constitutional and legislation levels. The Director for Environmental Policies and Norms at the DDF pointed out that the importance of keeping the elaboration of environmental norms as a federal function is due to the need for uniformity and standardisation 'because if it were local, there could be incoherence of norms in each of the federated states'. Nonetheless, this person argued that in the case of critical areas, such as the MZMC, there should be joint work between federal and local authorities. This is because in order to achieve air quality norms in Mexico City (where there are particular geographical characteristics and industrial locations) specific and individual emission norms must be stricter than in other areas like Veracruz (a coastal city) where the wind 'cleans-up' the atmosphere (see Chapter 2).

In relation to the administrative level, whereas federal authorities disagree on whether central government should participate or not in the implementation and enforcement of air pollution norms, local authorities seem to prefer having exclusive powers for the latter rather than jointly working with the federal government, unless the local authorities themselves require some kind of federal support (see Table 7.2). The interviews carried out with different federal and local authorities illustrate both sides of the argument in relation to whether the federal government should intervene in the implementation and enforcement of air pollution emissions for fixed sources (i.e. industrial emissions) in the MZMC. First, central authority views on fixed sources are divided. On the one hand, the head of the Office for Normative Inspection *(Subprocuraduría de Verificación Normativa)* at the Attorney General for Environmental Protection (PROFEPA) *Procuraduría Federal de Protección al Ambiente* - the federal agency for implementing and enforcing policies for fixed sources in the MZMC - said that from a legal point of view federal intervention in the MZMC is 'adequate enough' because the problem of polluted air embraces a huge variety of emission sources. Such emission sources include not only those under the DDF's responsibility, such as moving and non-industrial fixed sources, but also industrial fixed sources which require federal participation because of the complexities of regulation. On the other hand, the Attorney General at PROFEPA, pointed out that the industrial emission control special formula for the MZMC derived from the LGEEPA - whereby the PROFEPA is in charge of inspection and enforcement of regulations of all industries - should not continue. According to the attorney general, there is no reason why the MZMC should have a special regulatory system, compared to the rest of the federated entities where local authorities participate in industrial emission control when it is generated by local sources or in zones of local jurisdiction. It was pointed out that 'the intervention of the federal government...[in the MZMC]...in areas that could be very well covered by local authorities represents a centralist hallmark'. Additionally, the head at PROFEPA was not keen on the idea of getting the Federation involved when polluted air covers two or more federated entities, as in the case of the MZMC. The latter implies undermining the capacity of the federated states to sort out their own environmental conflicts.

Second, the local authority views in Mexico City seem to support the Attorney General's opinion at PROFEPA in the sense of getting powers for inspecting industrial emissions. For instance, the Director General for Environmental Issues at the DDF *(Dirección General de Ecología),* explained that there is 'no logical reason why the federal government has exclusive powers for controlling industrial emissions to the atmosphere...this is absurd'. According to this local authority, the underlying reason for this is due to the economic importance and vested interests that the industry located in the Valley of Mexico represents at the national level and thus the pressures on the Federation to keep control over it. This situation has hindered the DDF from properly carrying out the functions for which it is legally responsible. For example, although the DDF inspects industrial waste disposal, the pollution inventories each industry prepares (which include air, water, and so on) are usually given to the federal government and not to DDF authorities. Local authorities then have to ask the same industry for another emissions inventory, but only regarding waste disposal, which is an unnecessary duplicated task for government officials and the industry. While the DDF has already requested more powers with regard to inspection matters - something that PROFEPA has partially agreed to - the Director General for Environmental Issues confessed that in some ways, it is convenient for them not to have so many statutory powers with regard to inspection issues because of the lack of personnel and financial resources. However, this officer emphasised that the government of Mexico City should have these powers and that the institutional capacity for responding to new duties should not be a problem.

Overall, the interviewee opinions on federal and local government intervention and the existence of coordination agreements for environmental purposes - either by law or through regular informal meetings - confirm that a common government view for adopting an integrated approach for tackling air pollution in Mexico City exists. This requires central and local government participation, as well as the need to continue the process of devolution of power to Mexico City's local authorities which started in the late 1980s. During the mid-1990s, this process has included increasing powers for the Mexico City government at regulatory and administrative levels. Indeed, as from 1994, the ALDF can enact legislation on air quality issues, such as public transportation and traffic management (Constitution 1997; DDF 1993; EGDF 1994; Gamboa de Buen 1994: 158-159).

None the less, federal government intervention on air pollution issues at policy formulation and regulation levels is likely to continue for the rest of the century. The new 1996 air quality programme for the MZMC that enhances and updates the 1990 PICCA is a good example of continuous central intervention. According to a director general at SEMARNAP and in charge of the elaboration of the new air quality programme, the idea of such a new programme came from the federal government. Presumably, the principles of the new air quality management strategy were being discussed with other local authorities - particularly with the DDF and EdoMex - at the time of interviewing. However, the discussion and progress of the new air quality plan has failed to include, or at least inform, other members of the CMPCCAVM of the issues under debate. Although central government authorities announced the existence of this programme in the media during 1995, by the time the interviews were carried out with some members of the CMPCCAVM (July 1995), this body had not yet had its first meeting since the beginning of that year.

The results of the semi-structured interviews in this section reveal that heavy central government intervention and recent increased participation of local authorities has hindered other actors in society from participating in air quality management issues as recognised by the local governance approach. The existence of a blurring of responsibilities derives not from the various agencies - networks or partnerships - for managing air quality (as explained by the network approach), but from imprecise legislation (i.e. functional allocation) which can be seen in the division of responsibilities between central and local units.

The role of the Mexico City government

As seen in Chapter 4, the organisational changes that are taking place in the system of local government in Mexico City - i.e. direct election of a mayor at the upper level in 1997 and of local mayors at the lower level in the year 2000 - stem primarily from democratic and representative demands rather than from city management concerns. The current debate on local government organisation in this urban centre raises two major issues within Mexico City's urban environmental management in relation to air pollution. First, since democracy and representation are at the front of the political reform debate, it is important to query whether these two issues would make a positive impact on managing air quality. Whether democratisation may improve the functionality of this or any other city, and thus be reflected on urban environmental improvement, is a hotly debated issue with no definite or universal answer (see, for example, de Geus 1996: 194-197; Díaz Díaz & Perló Cohen 1994: 55-57; Doherty & de Geus 1996: 7-14; Ward 1990: 73-91). Scholar discourse and reality could be far apart, but at least, as Davey (1996: 65) pointed out 'in theory the extent to which urban governments are accountable to the local electorate should have a major impact on their effectiveness'. Second, democratic and accountability concerns have shifted attention from reviewing the advantages that the current system of government has for controlling air pollution. Some of these advantages relate to, for instance, the importance of DDF participation as a strategic body for coordination and standardisation purposes, as well as for liaising with other federal, local or non-governmental agencies as a single, city-wide authority. The positive aspects that can be identified from the way Mexico City has been governed and that should be taken into account when reforming local government, should not however, become an excuse for preventing the democratisation and accountability process in Mexico City at the upper and lower tiers of government.

The results of the interviews in Mexico City with various local authorities within the DDF suggest that the relationships of Mexico City's upper tier authorities with other agencies in terms of coordination have not been a major limitation for tackling air pollution. According to local authority views, there seems to be a low degree of tension between federal agencies and Mexico City's government, as well as between the latter and the 16 delegated units. This does not imply that there have not been some conflicts between central and local authorities when discussing the elaboration and application of air pollution regulations. Some of these conflicts have however, occurred due to overlapping of functions derived from imprecise legislation.

Although different views and interests from central government departments and the DDF arise when discussing actions for improving air quality in the MZMC, their relationship has been described as 'good' and 'friendly' by the local authorities themselves (see Table 7.3). There are two main reasons that may explain why Mexico City's government relationship with the central government is not often conflictive. First, whereas the creation of a Metropolitan Commission for Environmental Pollution for the MZMC (the CMPCCAVM) seems to have largely contributed in bringing all diverse governmental authorities (including the EdoMex) into one single forum of discussion where some degree of consensus has been reached, such a consensus has been centrally-led most of the time. According to the Director General for Environmental Projects at the DDF and member of the CMPCCAVM, the main function of the Metropolitan Commission has been to specifically act as the coordinating mechanism among all federal and local government agencies. This is in order to jointly formulate integrated actions against air pollution and implement them by involving all governmental sectors according to their own responsibilities. Not all local authorities however, believe that this metropolitan body has been efficient or effective enough, since central government imposes its views. For example, the Head of the Secretariat for the Environment at the DDF and also a member of the CMPCCAVM, explained that the current commission does not work properly 'because it is not an independent and autonomous body...it needs to have its own budget'. Second, the alleged lack of tension among federal and local authorities relationships has to do with party political issues and the lack of autonomous authorities in Mexico City, as well. Although a mayor's office could become powerful on its own, and relatively autonomous from the federal government, until 1997 the mayor had been appointed by the President of Mexico, and thus followed most of the presidential instructions (see, for example, Ward 1990 and 1998). Furthermore, the last three non-elected mayors involved in the implementation of the 1990 PICCA programme - i.e. Camacho 1988-93, Aguilera 1994, and Espinoza 1994-97 - all belonged to the same political party as the President: the PRI. As explained in Chapter 2 and 4, the initiative to adopt an air quality management strategy during the Salinas administration did not come from an autonomous mayor whose main concerns were oriented to air pollution, but from the President who regarded environmental pollution as one of the priorities on his political agenda.

Table 7.3 Local authority views (DDF) on Mexico City government's relationship with other agencies regarding air pollution issues

Upper-tier authority views with:	
the federal government	Relationship described as 'good' and 'friendly' Although different views arise among government bodies, consensus has been achieved No major coordination problems Conflicts arise due to overlapping functions from legislation
delegated units (lower tier)	Relationship described as 'good' and of 'much cooperation' No major coordination problems Positive and helpful participation of delegated units for carrying out DDF functions Delegated units should assist DDF in areas that are 'too local'
the EdoMex authorities (state and municipal)	Relationships described as 'good' No major coordination problems Occasional conflicts arise due to independent hallmarks of municipalities Although different views exist, consensus has been achieved
non-governmental organisations	Participation is welcomed Degree of influence depends on level of knowledge and type of proposals

Source: Author's semi-structured questionnaires

In the case of the relationship between the DDF (upper tier) and its 16 politically- and territorially-fragmented delegated units, the fact that the latter are non-autonomous entities whose functions derive from the mayor's office means they often follow what the mayor tells them to do. According to the Director General for Environmental Issues at the DDF, as all heads of the delegated units are appointees of the mayor and belong to the same political party as the mayor, there is no real opposition or discussion on most of the issues. There is actually more discussion on air quality issues between Mexico City local authorities and the surrounding municipal authorities (as well as with other non-governmental organisations) than with central government or delegated units. For instance, although the 17 conurbated municipalities to Mexico City are truly elected and autonomous authorities at the EdoMex, the first mayor of the Federal District during the Salinas administration had to get the cooperation and compliance from all of them in the application of the 1990 PICCA air pollution measures.

Regarding the relationship of Mexico City local authorities with other bodies, most interviewees at the DDF described this as 'good', depending on the type of the organisation. For example, when mentioning NGOs, the Director General for Environmental Projects at the DDF said that it is not possible to put all environmental organisations that participate at the CMPCCAVM in one single group because each has its own views on air pollution issues. These views correspond, either to the interests of the specific communities they represent (such *as Unión de Grupos Ambientalistas* who represent many ecological groups), or to international guidelines (such as Greenpeace). This person emphasised that, although unconditional support from environmental groups for governmental action is not expected, some of them only criticise governmental activity without proposing viable options. Nevertheless this interviewee recognised that some environmental groups make positive criticisms and suggest specific proposals:

> their degree of influence is variable depending on the level of knowledge they have and the will they show to participate in or outside the Metropolitan Commission...because the environment has become a citizen's priority, some groups use it as a political platform.

The fact that Mexico City's government involvement on air pollution issues has increased does not mean that all interested parties have also participated in the policy-making process. This situation is not only due to the lack of upper and lower democratic tiers of government, but also to a weak legislative body in the past and a non-democratically created metropolitan commission, the CMPCCAVM, for the MZMC, which have barely represented the interests of Mexico City inhabitants. Some examples can illustrate this. First, as already mentioned, Mexico City's local authorities that have participated in policy formulation regarding air quality issues have not represented the interests of Mexico City inhabitants since they were not elected. Second, at the time the legislative body for Mexico City ARDF was created (1988), the 1980s air pollution policies had already been discussed: the ARDF participation *in a posteriori* regulations such as the 1990 PICCA was legally minimal due to its fairly limited powers regarding environmental issues. As already seen the new legislative body ALDF can now regulate and enact environmental legislation for the Federal District but with no clear specification on its air pollution powers. Finally, although official publications (for example, the 1990 PICCA and the 1996 air quality programme) have mentioned that the strategy for dealing with air pollution has included the participation of national and international specialists, as well as environmental groups and citizens, the outcome of the interviews with environmental groups and academics in Mexico City tells a different story (see CMPCCAVM 1994a: I/17-I/19; DDF 1996; SEDESOL 1993 and 1994; STI 1990: 2).

Most of the non-governmental interviewees said that their participation, when permitted, did not influence the decision-making policy process. For instance, the Head of the Office for Environmental Pollution at UNAM and member of the CMPCCAVM said that during the Salinas administration, the state-owned oil industry PEMEX and the federal health authorities hide all available information on air pollution and related matters - such as health impact. Furthermore, according to my contact at UNAM they also denied the existence of damaging consequences as a result of the air quality situation in Mexico City. This interviewee emphasised that 'it is very difficult for politicians to accept that damaging effects exist because of air pollution: they do not like being told off...although the CMPCCAVM listens to us, most of the time they do not follow our recommendations'. Likewise, an interviewee at the National Institute for Public Health (INSP) *Instituto Nacional de Salud Pública*, said that the then Secretariat of Health (SSA) *Secretaría de Salubridad y Asistencia* during the Salinas administration was reluctant to take into account INSP research findings on human health. The new Secretariat for Health (Ssa) *Secretaría de Salud* under Zedillo's administration seems however to be more open and receptive to the work they have produced recently.

Other examples concern environmental groups. For instance, the head of Greenpeace Mexico's Atmospheric and Energy Campaign mentioned that, although there is more access to the Ssa and the National Institute for Ecology (INE) *Instituto Nacional de Ecología* under Zedillo, the DDF is still 'closed' to any outside environmental policy recommendation, especially regarding transportation issues. Another example is the president of the environmental group *Unión de Grupos Ambientalistas* and a member of the CMPCCAVM who also mentioned that their environmental suggestions were ignored, and, when taken into account, the main idea was distorted. For example, while this organisation launched the voluntary programme 'A Day Without a Car' in the mid-1980s, the programme was implemented by the federal and local governments after few years of twisting its original purpose. While such a programme was intended to be implemented on a voluntarily basis, the federal and local authorities made it obligatory. In addition, the president of this environmental group said that the policies to be discussed at the CMPCCAVM are usually previously agreed among governmental authorities. This interviewee insisted that 'from a citizen participation point of view regarding air pollution issues, there has not been a lot of progress'.

From the outcome of this section there are issues that fit better with the 'hierarchical' approach than with the 'market' or 'network' ones. First, the interviews reveal existing concerns on increasing delegated unit participation in air quality issues on a joint working basis, with the area-wide authority - the DDF. While the latter reinforces the results of the survey in Chapter 5, it also resembles the idea of creating a 'top-heavy' two-tier formula of government (as explained by the orthodox model), as some interviewees believe delegated units should assist the DDF in areas that are 'too local'. The empirical information does not however, seem to suggest with clarity which areas may be regarded as metropolitan and which as local. This situation is a classical example of what the orthodox and public choice approaches have identified as the functional allocation problem in metro schemes.

Second, Mexico City's government response to air pollution does not seem to show major problems as identified by the orthodox model regarding coordination and standardisation, financial equalisation, and equal distribution of diverse functions or subfunctions within the air quality management strategy. Nor has it shown major problems under public choice assumptions in connection with production costs (efficiency), transaction costs (governmental relationships), or externalities. This can be partly explained because the relationships between the DDF authorities and both central government and delegated units have not been too conflictual, and consensus has been achieved without major opposition. This does however follow on from the fact that central authorities have imposed their views on non-elected and centrally-appointed DDF authorities, and the lack of autonomous and elected lower-tier authorities. At the metropolitan level (i.e. within the MZMC), however, this situation has in turn revealed some problems of consensus building and restricted participation of other agencies and actors within society. This concern at the metro level raises the question of whether cooperation from the surrounding authorities of the EdoMex could be more easily obtained through an area-wide authority that would cover the whole metropolitan complex (as envisaged by orthodox views) or through other agencies which could serve as coordinating and consensus building, bodies networking with several participants (as envisaged by local governance views) (see also Table 7.3).

Third, the way in which air quality management policies are elaborated for the whole MZMC (i.e. through cooperative arrangements between the DDF and EdoMex authorities) raises the question of whether the transactional costs between Mexico City authorities and surrounding municipalities might be reduced with the existence of a metro authority for the whole metropolitan complex. As the current GDF has no control or power over independent and elected surrounding municipalities, the avoidance of externalities (positive or negative) depends on concerted actions between these local authorities. While the CMPCCAVM has served as an institutional channel or forum for discussion this specialised agency is not autonomous enough (i.e. it is financially weak). The existence of such an agency fits with the orthodox views of confronting political fragmentation through cooperation among local governments (in this case the GDF and EdoMex), with no structural changes in government. Under local governance assumptions, the CMPCCAVM has created some uncertainty and ambiguity over the way it operates and over who is responsible for what. At the same time, it has failed to ensure that other participants (non-governmental organisations or community groups) have contributed to the decision-making process. This situation, together with the predominant participation of central and local government units in Mexico City, reveals that the existing organisational arrangements do not properly fit with the assumptions of the 'network' approach which recognises the involvement of other networks or community groups.

Air quality monitoring systems

Monitoring air quality in Mexico City has been, for a long time, a federal function. It was not until 1992 that the former environmental secretariat SEDESOL gave DDF authorities the responsibility for all monitoring systems in the MZMC. Since 1986 the whole monitoring system - i.e. the Automatic Air Quality Monitoring Network (RAMA*) Red Automática de Monitoreo Atmosférico* - has been improved to such an extent that it is now regarded as one of the best air quality measurement systems in the world, compared to those in Los Angeles, New York and Tokyo (DDF 1994; WHO/UNEP 1992: 36). The air quality monitoring system in the MZMC currently consists of 32 real time or automatic, and 19 manual monitoring stations measuring different pollutants. These monitoring systems are located across the MZMC - the Federal District contains 21 real time ones and the conurbated areas of the State of Mexico, 11. The MZMC is divided up into five main areas that continuously report on the air quality situation for the whole metropolitan zone (CMPCCAVM 1994a II/50-II/54 and 1994c: 34-39; SEDESOL 1993: 156; STI 1990: 29-39). During the fieldwork period in Mexico City, several visits were made to monitoring stations in various delegated units across the MZMC where discussion took place with some of the technicians that are in charge of the operation of the RAMA system. Other visits to monitoring stations were also made to autonomous organisations such as the National Autonomous University of Mexico (UNAM) and the Mexican Institute for Petroleoum (IMP). The results that stemmed from the latter showed that the monitoring system is fairly sophisticated - especially if compared to that of London. The effectiveness and efficiency of the system for measuring pollutants and providing continuous information on air quality standards proved to be adequate enough in terms of coordination, uniformity and standardisation.

The success of the monitoring network in Mexico City has to do not only with technological issues, but also with the fact that there has been only one organisation (either federal SEDUE, SEDESOL, SEMARNAP or local DDF) that has been in charge of managing and operating the whole system. First, according to the local authority in charge of the automatic real time monitoring system at the DDF, the type of equipment and the whole system is one of the best at international level, due to its continuous calibration and standardised method of measurement. The calibration of the equipment is certified by three international agencies, the United States Environmental Protection Agency (USEPA) every six months, the German agency (GTZ) every three months, and by the Japanese International Cooperation Agency (JICA) every two months. All pollutants within the automatic network are monitored by consistent methods, for example, ozone is measured by U.V. Photometry. Second, the management of the monitoring network by one single authority has proved to be efficient because standardised information on the MZMC-wide air quality situation can be obtained by the same authority at once, and at every hour, for most pollutants. All the information sent by the 32 monitoring stations across the MZMC is analysed through a complex but uniform computer system at the DDF which provides accurate measurements of pollutants that are, in turn, sent by fax or computer through e-mail to the federal and local authorities that are in charge of implementing the contingency plan at the DDF if high levels of pollution are reported, i.e. if the permissible levels of the Mexican value index IMECA are exceeded.

The interviews with federal and local authorities and non-governmental organisations show no major criticisms of the way in which the air quality monitoring system has operated or the fact that it is no longer managed by the federal government or the delegated authorities. On the contrary, as the Director General for Environmental Projects at the DDF emphasised, 'as is true that some tasks need to be done by the *delegaciones*, such as getting local communities more involved in air pollution issues, it is impossible to decentralise the automatic air quality network. This is simply because it would imply that every delegated unit would have to take care of every single monitoring station within its jurisdiction...something that is not acceptable'.

There are two aspects, however, that need to be addressed so as to keep the whole system working properly and to make it more accessible to the public: both involve issues relevant to Mexico City's local government reorganisation. The first has to do with transfer of the monitoring responsibilities from the SEDESOL to the DDF in 1992. Such a transfer had a negative impact on the training programme given by the USEPA to Mexican technicians in relation to the use and maintenance of air quality equipment. My contact at the Automatic Air Quality Monitoring Network at the DDF, who experienced the financial transfer, pointed out that as soon as the air quality monitoring functions became a local responsibility, USEPA cut down its financial assistance for this purpose. He commented that currently there are not many specialists who are trained in using and maintaining this equipment; there are only eight people currently working at the RAMA Office and, according to him, this number should be increased. This interviewee said that if further devolution of power to local government was going to take place, it should be accompanied by financial and technical assistance in order to avoid a lack of continuity in the implementation of certain policies or programmes. In this case, the federal government should have ensured the continuation of this training programme.

A second aspect that requires some serious consideration is that of public accessibility to the information provided by the RAMA Office. According to the outcome of the interviews with local authorities at the DDF, there seems to be no distribution of the air pollutant monitoring results to the delegated units in Mexico City. This particular situation - confirmed by the local authorities at the delegated units - prevents local communities in Mexico City from getting details on the air quality situation in the MZMC. Although access to the RAMA Office for the results of the air quality situation is not 'blocked' to anyone who requests it - as personally experienced - a single office at the upper level of government distributing them does not seem to be close and accessible enough for a population of more than eight million people in Mexico City, let alone for the whole MZMC. If one of the criteria for an adequate air quality system is that information should be continuous and accessible so as to alert the population to a pollution episode, then the intervention of the delegated units for information purposes could be extremely helpful. While the media play a major role in this, especially when a contingency plan needs to be implemented, there are vulnerable groups such as asthmatics, the elderly, people with cardiovascular or respiratory problems, who regardless of a contingency situation, need to have information on pollution levels.

As confirmed by some people at the RAMA Office, although the costs for installing monitoring equipment is extremely high, the access to the information on the air quality situation is not, and yet could easily be reached by computer. The requirements for obtaining the information include the acquisition of a compatible PC computer (386 processor; 40 mb in the hard disk; 2mb in the memory), a modem, and a telephone line to call the DDF, which altogether are not expensive and are easy to get on the market. Once installed, the RAMA Office provides technical assistance for the operation of the system (see DDF 1994). The installment of 16 computers, each in one of the 16 delegated units, would certainly not represent an unthinkable expense for local budgeting, but would assist the DDF in its task of informing the population, especially the vulnerable groups whose interest is greater than that of the rest of the population. The media, as in London, seem to report episodes only when the levels of pollution are extremely high (see also the survey results in Chapter 5).

The public good dimension that the public choice model inserts into the discussion for reorganising local government structures helps to fully understand, paradoxically, the advantages of an area-wide authority for managing air quality monitoring systems. The results in this section show that the system managed by one single authority has proved to be coordinated, efficient, and equally distributed. Transactions costs of the relationships of different participants involved in air quality management are reduced because information on the whole MZMC air quality situation can be obtained from only one authority. Additionally, air quality information is equally distributed (it covers the whole metropolitan complex, with standardised information) and efficiently (with reduced disseminating costs) by fax or computer (e.g. e-mail). It is important to note that the metropolitan coverage of the monitoring system was made possible through the financial equalisation that central government provided until 1992. This situation raises the question of whether the enhancement and maintenance of the whole monitoring system will only take place within boundaries of Mexico City as the GDF does not have authority over surrounding municipalities. It is then worth considering the need for an area-wide organisation (larger than the GDF) to ensure that the system will continue operating as it has so far.

Transportation and traffic management

As seen in Chapter 2, among the sources of air pollution in Mexico City, those that relate to oil fuel consumption are the most significant in terms of weight and toxicity. While industry remains as the prime polluter for such pollutants as SO_2, road transport is the main polluter for other pollutants such as CO, NO_x, and HC. Again, while private cars are the main source of pollution, they account for only 21.4% of the journeys in the Metropolitan Zone of Mexico City, by mode of transport. All other journeys are distributed among other means of public transport such as buses, underground, trolley buses and light trains (see Table 7.4).

Table 7.4 Journeys in Mexico City by mode of transport

Mode of transport	*Distribution of journeys (%)**
Cars	21.4
Buses	3.8
Underground	3.2
Minibuses	42.0
Taxi	3.8
Other (light trains, motorbikes, trolley buses	5.8

* Overall percentage (100%) is not available

Source: DDF (1995)

Since moving sources are regarded as the main cause of the problem most government strategies and non-governmental proposals have been oriented to prevent and control road transport emissions (see, for example, Bravo Alvarez 1993: 7; Calvillo 1993: 3-6 and 1995; CMPCCAVM 1994a: II/12-II/27 and 1994c: 26-27, 65-69; DDF 1995: Greenpeace 1993: 20-23; Quadri 1993: 10; SEDESOL 1993: 158-164 and 1994: 223-232; STI 1990:31). The government response to this problem has demanded considerable transport planning and traffic management restrictions in a city where for many years the use of the private vehicle has been favoured. An insufficient and corrupt public transport system, plus the new environmental demands of the 1990s, has led the Mexico City government to produce two major, integrated plans for the city (one in conjunction with federal authorities) to re-arrange and considerably increase the use of public transport but not without continuing to favour private vehicles. These two plans are the 1995 public transport and traffic management programme *Programa Integral de Transporte y Vialidad 1995-2000*, and the 1996 air quality programme (Proaire) for the Valley of Mexico jointly produced with federal agencies *Programa para Mejorar la Calidad del Aire en el Valle de México 1995-2000* (see DDF 1995 and 1996).

Diverse interviews with federal and local authorities in Mexico City showed the importance of the control of public transport systems and diverse traffic management issues by an upper tier of government (such as the DDF). The latter sheds light on the issues of institutional coordination among different authorities for the whole MZMC, including municipalities and delegated units. The issue of social participation in transport policy formulation however, was severely criticised by environmental groups that pointed out that their proposals have simply been ignored by the government.

Local authority views on transportation systems and traffic management issues are that these have to be controlled at metropolitan level, and if they are not, at least at Mexico City government level. First, the Director General for Environmental Projects at the DDF said that the reason why a metropolitan approach is required for road transport is because it is very important to have 'a detailed analysis of the impact that an action at local level [lower tier] could have over its surrounding areas. For example, a head of a delegated unit might like to divert traffic from a conflictive junction within its politico-administrative boundaries, but because this may produce collateral effects, creating additional traffic problems in the surrounding areas, a strategic authority such as the DDF should intervene to solve the problem'.

Second, when addressing the issue of coordination among several units, the interviews show that there seem to be no major coordinating problems between the authorities at the DDF and the authorities at the delegated units which help to implement all traffic management policies. As explained before however, this has very much to do with the fact that all delegated authorities are subordinated and report to the mayor. However, the relationships between DDF authorities and the other local authorities of the surrounding municipalities have produced some conflicts when enforcing policy issues regarding transportation and traffic management regulations. For example, the Director General for Ecology at the DDF and member of the CMPCCAVM, explained that disobedience at lower tier authorities is less common within the Federal District boundaries, than with some authorities at the municipal level in the MZMC. This person said that 'Municipal Presidents, as opposed to the heads of the delegated units, are elected and thus sometimes like to show their independent status. So, they do not follow or implement some policies that are for the whole metropolis and not only for the municipality - this happens even if they are from the same political party as the

Governor [of the federated state EdoMex]'. By way of illustration, when the programme 'A Day Without a Car' was launched for the whole MZMC, the Municipal President of Cuautitlán Izcalli in the federated state EdoMex told the inhabitants within its territorial jurisdiction that the programme would not be enforced in that municipality.[1] This, despite the fact that the predominant winds specifically blow from the north to the centre of Mexico City transporting air pollutants emitted there. The reason given by that local authority was because the municipality was too far away from the affected or polluted area (see Map 4.2). According to this officer, this decision was clearly oriented to politico-electoral purposes as well as due to a lack of knowledge and strategic view of the problem. The decision taken by the Municipal President, though, was overturned and the programme started operating fully in the surrounding municipality of Cuautitlán Izcalli.

Third, although some federal and local authorities explained that future elected lower tier authorities in Mexico City may prolong policy discussion or hinder the enforcement of policies because of politico-electoral reasons, the results of the fieldwork show that the 1990 PICCA programme has been enforced across the whole MZMC. This situation, which has included the participation of elected municipalities of the surrounding federated state EdoMex, proves that democracy and increased autonomy of lower tier authorities (i.e. delegated units) may not be obstacles for implementing transportation and traffic management policies in a coordinated way. As the Director for Environmental Policies and Norms at DDF pointed out:

> ...ideally, democracy will necessarily lead to more efficient public administration...democracy will always guarantee a better government, and a better response from civil servants because if their response is not adequate, then voters can elect someone else. Thus, officers feel more committed to citizens in the delivery of services, implementation of actions, and so on.

Nevertheless, this officer emphasised that by democratising Mexico City, some problems are likely to emerge given the politico-economic situation in the country in the short term, but not necessarily in the medium or longer term. If democracy and accountability do not seem to prevent coordinated and effective action but to foster it, then concerted actions among all DDF and delegated unit authorities become extremely important in order to achieve consensus and support from other federal or local elected authorities.

Most criticisms of public transport and traffic management policies by non-governmental organisations are either because the programmes themselves have lacked an integral vision for transport planning and traffic management, because corruption of local authorities has impeded the enforcement of many of them, or simply because society has not observed legal measures. Such is the case for the 'A Day Without a Car' programme where motorists preferred bribing police officers than respecting environmental regulations. On the one hand, while statutory powers given to police forces were necessary in order to stop cars for being on the road on a prohibited working day or for ostensibly polluting, their traffic management responsibilities have not only taken time away from carrying out other security and criminal functions, but have not been properly implemented. On the other hand, the fact that Mexico City inhabitants have bought a second-hand car since the implementation of 'A Day Without a Car' plan, participated in bribing police, as well as paid for emission control test number plates without having a proper vehicle test, indicates how society has reacted to such a plan. Civil disobedience, as explained by federal and local authorities (including NGOs), stems from a mixed and complex set of issues. These involve economic and politico-electorate issues (e.g. public resistance to comply with governmental mandates), the way in which private vehicles have impacted Mexican society, the lack of environmental awareness, and the need for an adequate and suitable public transport system. The latter corresponds with the results of the general survey carried out with local authorities in Mexico City (see Chapter 5).

The continuous criticism that environmental groups have for governmental actions regarding transportation systems and traffic management is derived from the way the government has responded to their demands and proposals. So, for example, an environmental campaigner for the environment, explained that the enlargement of the motorways within and around Mexico City's boundaries seemed to be linked to the embedded political and economic interests of present and past government authorities (central and local) which have shown no concern regarding environmental or local community issues. Presumably, since the building operators are owned by officers from the last two administrations, a small group of ex and current officers would financially benefit from these huge public works at the expense of Mexico City's air quality situation.

Public transport and traffic management issues in Mexico City constitute good examples that are relevant to the hierarchical and public choice approaches. This is because the various governmental agreements that exist within Mexico City and the MZMC are oriented to avoid externalities within the whole metropolitan area. The amount of coordination and consensus building on the past of all authorities involved in diverse traffic management issues - for example banning cars one day a week - is also revealed in this section. It is of significant importance to highlight that the same concerns raised by local authorities in London about closing roads without consulting other local authorities and about the need to have a metropolitan approach in these matters in order to avoid spillovers, was also raised by the local authorities at the DDF. The conflicts that have arisen between DDF authorities and surrounding municipalities in the implementation of some traffic management issues again raise the issue of considering the need for a larger area-wide authority to deal with these matters within the whole metropolitan complex.

Emergency plan

The application of the contingency plan in Mexico City *Plan de Contingencias Ambientales* when critical episodes of air pollution occur, proves how relevant the existence of an upper tier of government can be for coordination purposes. Whether the plan has been seen to accomplish its main objectives, and thus contribute to better air quality standards in the MZMC seems to be a more debatable issue. The way in which official publications describe how the emergency plan is implemented leads one to believe that the governmental response in case of an episode is quick and well coordinated among public and private bodies.

The application of the plan is initiated when the CMPCCAVM gets the information from the RAMA Office that the permissible levels of the index value IMECA have been exceeded. Once the CMPCCAVM decides to implement it according to pollution levels and meteorological conditions, the emergency alert is distributed in a 'cascade' system to all federal and local authorities in the DDF and the federated state EdoMex to comply with the regulations. Originally, the plan was divided into three phases. Phase 1 (called at 250 IMECA points) required a 30-40% cut in industrial emissions, a 50% reduction in the use of government vehicles, the halting of street repairs, and drivers were requested not to use their cars. Phase 2 (called at 350 IMECA points) required 50-75% cut in industrial emissions, schools were closed, and the 'A Day without a Car' programme was extended to two days a week. Phase 3 (called at 450 IMECA points) basically required all industries to stop their activities and the banning of the operation of any source of pollution. In practice though, these measures have been interchangeably implemented during Phases 1 and 2 (Phase 3 has never been activated) and have not always been enforced - especially regarding industry. At the time of writing, there are government discussions on whether calling an emergency situation (Phase 1) should be lowered to 200 IMECA points (see Barrera Echeverri 1993: 42-46; CMPCCAVM 1992: 16-20 and 1995a: 85-86; DDF 1996: 232-237; Elsom 1996: 203; SEDESOL 1993: 166-167; 219-222).

In order to find out if the governmental description was accurate, the interviews sought the authorities' views on the functionality of the plan. The findings of the fieldwork in Mexico City confirmed that the level of coordination within federal and local authorities and the speed at which information is distributed across the MZMC when a contingency exists, are fairly high. Indeed, the interviewees at the RAMA Office at the DDF said that the continuous information that the RAMA monitoring network provides is sent to the local authorities directly responsible for the implementation of the contingency plan. The time for informing authorities of the existence of a contingency and thus for the application of the emergency plan is between half an hour to one hour. The analysis of a possible emergency situation is made by central and local authorities between twelve and two o'clock in the afternoon; a period of two hours is what these authorities take for identifying a possible contingency occurrence.

Diverse federal authorities corroborated the speed at which information is distributed to them so as to implement the contingency plan according to their responsibilities. One of my contacts at PROFEPA, for example, explained that the Secretary General at CMPCCAVM (headed by the Secretariat for the Environment at the DDF at the time of interviewing) usually let them know about a contingency situation at two o'clock. In turn, PROFEPA notifies its inspectors (who carry portable radios) about the contingency and ask them to put aside any other activities but those related to verifying that industries are complying with a reduction in their productive processes depending on the phase of the plan. The industries that signed an agreement to cooperate with the implementation of the emergency plan are usually notified by the

CMPCCAVM or by PROFEPA by e-mail, fax or phone. According to this person at PROFEPA, it takes around half an hour to one hour for all inspectors concerned to start their industrial visits, once the CMPCCAVM have notified them about the contingency. The effectiveness of PROFEPA inspections during the emergency alert, however, depends on the contingency duration. Another officer at PROFEPA pointed out that 'if the contingency lasts only 12 hours, it is very difficult for PROFEPA to visit the corresponding industries in all the MZMC; but if it lasts longer, then the required inspections are more likely to happen'.

Another example that shows the speed at which federal agencies get to know about an emergency situation is that given by the Director General of Environmental Health *Dirección General de Salud Ambiental* at the Ssa. This person explained that they usually receive the notification of a contingency situation from the DDF by phone. As soon as they get the information, they pass it by fax or phone to other health organisations, so they can act accordingly. The interview carried out with this officer took place late in the afternoon on 31st July 1995, on the same day as there was an emergency situation. According to this interviewee, that day, the contingency started at one o'clock in the afternoon and by twenty past one, they already knew about it. By twenty to two in the afternoon they had already sent messages to all health authorities across the MZMC.

The importance of one single wide-authority for Mexico City which receives continuous information on levels of air pollution for the MZMC, and which is in charge of disseminating the information to other public and private authorities when an emergency situation exists, is clear. If the adequate implementation of an emergency plan involves several activities across a huge metropolitan area like the MZMC (such as facilitating traffic speed, interruption of pothole activities and gardening and irrigation on traffic roads, and informing the population) the participation of lower tier authorities may be even more helpful in terms of speed and geographical coverage of local areas.

Despite the emergency plan for the MZMC seeming to be a well coordinated action, it has been criticised by academics and NGOs for the criteria being used to initiate the plan. Criticisms have focused on the fact that the threshold limits of the IMECA index value are too high compared to international values (see Chapter 2). The outcome of the interviews showed a variety of answers on whether the plan has worked or not. While local authorities at the DDF agree that the plan has worked because it has brought down high peaks of ozone as well as informing the population about the emergency situation, federal authorities contradict themselves on its functionality. For instance, while one of my contacts at PROFEPA believes the plan works adequately in terms of bringing down the pollutants and for educational purposes, another contact at SEMARNAP and at Ssa have pointed out that the IMECA index value should be less tolerant. That is to say, the contingency plan for the case of Ozone should start when pollution levels reach 200 IMECA points (or even before) and not when they reach 250 IMECA. Environmental groups and academics joined these two interviewees in criticising the inadequacy of the IMECA criterion values in accordance with to international standards and human health purposes. It must be noted however, that the contingency plan has successfully operated in terms of reducing car use, decreasing industrial activity, and making people aware of the air quality situation. It has to be emphasised that, as an environmental campaigner at Greenpeace said, 'whatever the actions are, they must be preventive and not aimed at solving the problem'.

The case for the implementation of the contingency plan in Mexico City as presented here reveals the importance of ensuring a coordinated and equal governmental (and non-governmental) response to the problem, as envisaged by the orthodox model. The latter can be clearly seen when an emergency alert is officially launched through a 'cascade' system initiated by the CMPCCAVM. The successful implementation of all measures in such a plan needs the equal and quick cooperation of all political units in a metropolitan complex in order to avoid the externalities that may be created by some units, which are not willing to take part as they do not suffer from the pollution that is originated either within their jurisdiction, or in other areas of the metropolitan area.

Conclusion

The outcome of the semi-structured interviews carried out in Mexico City suggests that the existence of one single city-wide authority for the whole DF - the DDF - has facilitated the government response to air quality management across the MZMC. This area-wide body has allowed a coordinated and strategic approach for managing air quality issues, particularly regarding monitoring systems, transport and traffic management, and the implementation of the emergency plan. The advantages of this situation, though, have been underpinned by the fact that the DDF has not been democratically elected and thus has not represented either Mexico City inhabitants' or NGOs' views. While the devolution of power from the federal to the local government is required, improving air quality could be maintained and fostered in a more democratic system. Such increase in local authority air pollution responsibilities however, needs to be accompanied, either by financial and statutory powers, or by the possibility of Mexico City authorities signing coordination agreements with the federal government for financial, technical or personnel assistance. Additionally, clearer legislation on the distribution of functions in this regard would facilitate the implementation and enforcement of policies by the federal and local government.

Although heavy intervention by federal government agencies continues, the participation of local authorities in air quality management issues - the DDF and EdoMex - has increased. Yet, central and local authorities have failed to create the institutional mechanisms to make community participation possible and accessible during the process of policy discussion for improving air quality. On the one hand, while the created metropolitan commission - the CMPCCAVM - has not always achieved concerted action among all government participants, it has often ignored non-governmental and academic proposals on improving air quality. On the other hand, while the political reform process on reorganising the structures of local government in Mexico City have addressed the importance of citizen participation by means of creating democratically elected units of government, it has failed to include air quality management concerns in the debate. At the time of interviewing, there was no discussion either on enlarging the scale of the current GDF government unit, or on creating a metropolitan authority for the whole MZMC for managing diverse air quality issues. Furthermore, the findings did not show which functions within the current air quality management strategy can actually be allocated to the delegated units in Mexico City when they become elected authorities. The participation of lower tier authorities seemed adequate only for disseminating air quality information in their own local areas as well as for channel citizen demands to upper levels of government.

Notes

1 The legally-binding 'A Day Without a Car' programme *Programa Hoy no Circula*, consists of banning vehicles one working day a week according to the colour displayed on each vehicle's number plate. This programme, which initiated in November 1989, has been recently modified by the new 1996 air quality plan (see DDF 1996). The new measures allow newer cars (which have been produced since the year 1993) to be exempt from the banning regulations, only if they pass a vehicle emission test.

8 Conclusions and Further Avenues for Research

There are two main responsibilities for big city mayors. One is to be a political leader and to represent the city ideals and dreams for a better future. The second is to be a good manager of technical solutions to city problems...

Jaime Ravinet
(current Mayor of Santiago de Chile)

Introduction

This research was written with the objective of contributing to the understanding of local government studies as well as of air quality policy research. By adopting a comparative approach which looked at two case-studies (London and Mexico City), the book sought to address and analyse how local government operates in relation to air quality management, considering this in the light of diverse models of local government. The first four chapters (Chapters 1-4) provide a descriptive and analytical insight into air quality management issues and diverse approaches of local government - including a description of the current local structural arrangements in both case-studies. The three following chapters (Chapters 5-8) then seek to examine central and local authority views, as well as those of other key policy actors on the organisation of local government in both urban centres - London and Mexico City - in relation to air pollution control.

The aim of this final chapter is not to repeat the conclusions presented at the end of each of the chapters but to draw them together. The first section draws together the main differences and similarities between the two case studies. The second section looks at the applications of the models of local government to air quality management (as discussed in Chapter 3) and re-examines them in the light of the empirical material. The last section highlights the most relevant findings in the book and includes a brief section on further avenues for research.

Main findings: a comparison

The findings of the two case-studies examined in this book resulted in a series of similarities and differences. First, when exploring air quality regimes either on a single or comparative basis, attention needs to be paid to how society and government perceive and look at air pollution in a particular urban location. Determining whether or not poor air quality exists, and thus what levels of pollution are permissible (commonly through domestic air quality bands and guidelines), is at least in part the result of social and political constructions. Indeed, levels of air pollution, types of air pollutants, and adverse effects on the health and welfare of human beings and the environment can differ considerably from city to city according to scientific data, and still be considered issues of much concern: just as happens in the cases of London and Mexico City. So, for example, whereas in London concentrations of diverse air pollutants have only a few times breached the WHO health guidelines and vis à vis Mexico City, have regularly exceeded those threshold limits, air pollution is regarded as an urban environmental priority in both cities. The fact that two cities as distinct as London and Mexico City suffer from perceived poor air quality (albeit the scale of the problem is different in each case) shows that some interconnection between scientific evidence and social judgements in urban centres may exist, irrespective of wealth or geographical location, in either a developed or developing country.

Second, managing air quality in London and Mexico City has been largely the result of the way in which the UK and Mexican central governments have respectively addressed and responded to it. Similarities and differences in this regard are however, linked to a distinct historical process in the evolution of air pollution in each country. The UK's impressive record to recognise and respond to environmental degradation (e.g. by creating the first world's environmental agency in 1863 and enacting the world's first comprehensive air pollution control Act in 1956) can be seen as a logical outcome of a process of industrialisation that took place first in Britain. While during the 1950s and 1960s levels of traditional air pollutants started to decline in London, they started to increase in Mexico City as soon as industrial development began to rapidly grow within and outside the city during these two decades (see Chapter 2). However, the 1970s represent for both cities the beginning of rising levels of air pollution from a similar source: road transport emissions.

What it is important to highlight during these last three decades of UK and Mexican government policy response, is that although high levels of pollution started to increase in these two urban centres during the 1970s and 1980s, both national governments were slow to react and showed environmental lethargy, apathy or ignorance in their successive administrations. Whereas the UK government had a piecemeal, incoherent and incremental response to air pollution, the Mexican government had a heavily centralised, non-democratic and specific, or sectorial, response to it. In any case, both situations led the UK and Mexico into a confused jungle of institutions and laws regarding air quality management issues. It was not until the 1990s that a more integrated response to environmental problems, and thus to air pollution, took place in both countries. This time the Mexican government first officially recognised air pollution as an environmental priority in 1986 and launched a strict and integrated plan in 1990 to control air pollution through the PICCA (which was enhanced and updated in 1996). By contrast, the British government put air pollution up with its highest priorities on its public policy agenda only in 1994 and, few years later, laid down the principles of a more coherent air quality management strategy in 1997 through the NAQS (see, for more details, Brañes 1994: 83-85, 117-118; DoE 1996 and 1997; González Márquez & Cancino Aguilar 1994: 36-37; Liverman 1993: 228; McCormick 1991: 9-10; O'Riordan & Weale 1990: 12).

Third, the main source of air pollution in London and Mexico City is road transport emissions, particularly from private vehicles. Although industry is still a significant source of pollution in the case of Mexico City, the main pollution control policies in both cities need to be oriented to the improvement of public transport systems and stricter traffic management regulations. As car ownership and car use continue to grow in both urban centres however, alternative traffic management measures and improved public transport may be offset by increasing levels of road transport emissions. Interestingly, while the number of cars continues to grow rapidly in London and Mexico City, the design of the air quality management strategies in both cities are largely influenced by the own UK and Mexican government's long-standing commitments to road building, and a relatively long period of under-investment in public transport. The widening of London's orbital motorway (the M25) to 14 lanes along its busiest roads and the expansion of Mexico City's metropolitan orbital motorway which surrounds the MZMC, exemplify the importance road building programmes have for these two countries. As with London's

M25, the new MZMC orbital motorway is intended to serve long distance traffic and reduce congestion problems in the metropolitan area. It is quite likely however, that the latter will generate, for example whenever new roads are built, more traffic and thus increase environmental pressure on the Valley of Mexico City. The take-over by a Labour government in the UK raises the question of whether there will be a change in London post 1997 regarding road building schemes or not; similarly, the democratic process that Mexico City's system of local government is undergoing (e.g. directly elected authorities) raises the question of whether major road building projects will continue or not.

Fourth, the capacity of the existing air quality monitoring networks is different in each case. While both networks comprise manual and real time monitoring equipment, Mexico's automatic system is much more advanced and sophisticated than the one that exists in London. Despite this distinction, both monitoring networks have been subjected to some criticism due to the particular location of their air quality monitoring stations. Apart from this, the London system has also been criticised due to the limited number of real time monitoring sites and the lack of a complete, comprehensive and standardised air quality database. An adequate air quality monitoring network (like the one that operates in Mexico City) requires not only comprehensive and standardised databases, but a coordinated system which can provide continuous and accurate information on diverse pollutants. While there is not a single rule to determine the number and location of monitoring sites, these should, at least, cover, and be evenly situated, across the whole metropolitan area.

One final point that arises from this comparative study is that while national air quality standards are stricter in London than in Mexico City for several pollutants, there are similar concerns on the way in which air pollution concentration is banded in both cities. In the case of London, two of the most relevant criticisms of the current DETR air quality criteria (banded as 'very good', 'good', 'poor', and 'very poor') are that it is too wide and vague and does not account for mixtures of pollutants. Likewise, in the case of Mexico City, criticisms of the MZMC air quality criteria (banded as IMECA: each pollutant equal to 100 points) indicated that the bands are too wide and do not take into account all the synergetic effects of all mixtures of pollutants. While air quality bands have been criticised in both urban centres for being too tolerant, which in turn, makes pollution episodes less likely to occur, an immediate response to high levels of air pollution through an emergency programme (i.e. contingency plan) has only been developed in Mexico City, as such episodes have recurrently breached Mexican air quality guidelines.

Table 8.1 Local government organisation in London and Mexico City by the year 2000

	London (as Greater London)	*Mexico City (as the Federal District)*
System of local government	- NO legislative body NO supreme court	Executive: Mayor + GDF Legislative: ALDF Judiciary: TSJDF
Tiers of local government	**Upper tier:** Greater London Authority (GLA): Mayor + Assembly **Lower tier:** 33 local units (32 boroughs and the City of London)	**Upper tier:** Government of the DF (GDF): Mayor + Centralised, Deconcentrated and Parastatal units **Lower tier:** 16 local units (16 delegated units)*
Democratic system	**Direct elections:** Mayor + Assembly + all 33 local units (1 year in 4)	**Direct elections:** Mayor (every 6 years) ALDF and all 16 local units (every 3 years)

* Granting full municipal status to delegated units is still under discussion

Source: Various

Before turning to the application of local government models in air quality management, it is important to make some concluding comments regarding the structures of local government in London and Mexico City. As seen in Chapter 4, the two urban centres differ from each other in a number of organisational respects (see also Table 8.1). Both case studies, however, provide a substantial basis for making some useful generalisations about the formation and development of two distinct local government regimes during the 1990s. The common elements found in this research can be summarised in two main points. First, in both cases continuous pressure for governmental changes exists. While such changes are certainly being influenced by the specific and embedded historical characteristics of each city, they seem to be equally oriented to strengthen local authorities by means of creating more democratic units of local government. Second, the idea of creating and/or democratising local units of government is leading to the adoption, by the turn of the millennium, of a variant of a metropolitan scheme as discussed in Chapter 3. In London, the concept of a metropolitan authority is reproduced with the creation of the GLA which will jointly operate with the boroughs in a number of functions. Although this new area-wide unit will not be the kind of large-scale authority that existed in the past with lots of staff and powers (e.g. former LCC and GLC), it will be a metropolitan strategic authority. In Mexico City, the new local government reorganisation (i.e. the possibility of directly electing the mayor and the local mayors of the 16 delegated units, both with a range of powers) resembles a variant of a 'top-heavy' metropolitan government. The process of structural reorganisation in London and Mexico City that has taken place over the last decade has been accompanied with some devolution of power from central to local level (both at the upper and lower tiers), but it has not always included financial support mechanisms.

The next section examines the operation of local government in relation to air quality management. This is considered in terms of the models of local government as explained in Chapter 3 and re-examined in the light of the empirical material. It is important to acknowledge that the discussion on reforming local government in London has been invigorated by air quality concerns; by contrast there has been a lack of debate in the case of Mexico City.

Corollary: a metro scheme for improving air quality?

In some respects, the picture that emerged from the two cases studies of how local government operates in relation to air quality management is fairly consistent with some organisational assumptions that are found in the literature on local government. The empirical work brings into light the advantages and disadvantages that the organisation of local government in each city has for dealing with air pollution as well as some of the lessons that can be learned for best governmental practice. It is clear that the empirical material showed the need for an area-wide coordinating authority for managing air quality with some joint participation of lower-tier authorities. Hence, the results in both case-studies suggest that contemporary local government practice fits better with the 'hierarchical' than the 'market' or 'network' approaches. However, in the case of London (and to a much lesser extent with that of Mexico City) there are some issues that can also be associated with the local governance approach. This section thus presents some of the most significant aspects of the empirical material in connection with the orthodox and local governance approaches and concludes by commenting on the relevance (advantages and disadvantages) to this and future research on air quality of the three models as outlined in Chapter 3.

In London, the current system of local government has implied institutional fragmentation, strong central government intervention, public-private partnerships and private networks when dealing with air pollution. The findings of the interviews suggest that such a system of political fragmentation at the local government level has created a number of metropolitan problems or concerns as identified by both the traditional orthodox and the local governance perspectives. First, conventional thinking argues that a fragmented system of local government can create problems of equity and financial equalisation among diverse local units. It seems that these two problems are manifested among the London boroughs in relation to diverse components of the air quality management system which have broad catchment areas. This is because some boroughs have had more financial resources, or simply, because air pollution has been regarded as a priority by the local authority. A clear example of this is the type of monitoring equipment that is managed by each borough. Being one of the leading boroughs in London for local air quality management initiatives, the City of Westminster, for example, has both the political interest and the resources to locally run its own expensive but fairly accurate monitoring system - OPSIS. At the time of interviewing very few boroughs actually had their own automatic equipment. Although one of the main reasons why boroughs would not buy this equipment was the lack of statutory powers regarding monitoring, now that they have such powers they still do not have enough resources for acquiring such equipment. As the central government has failed to provide financial support on this matter (leaving the local monitoring system as a non-harmonised network), an area-wide body according to traditional assumptions would be able to shoulder local authority expenses on air quality monitoring stations. The new metropolitan authority for London - the GLA - may solve some of the problems of promoting equity and financial equalisation across London (for monitoring and other broad catchment subfunctions) as long as it proves to be a politically impartial body. This is important simply because if the GLA benefits only those boroughs which are their political allies and/or bows to political pressure from the central government to do so, then equity and financial equalisation will not be obtainable.

Second, traditional orthodox views assume that the issues of community and lifestyles are involved in determining which issues have a metro or localised character. The outcome of the fieldwork in London seems to corroborate the fact that air pollution is basically an area-wide issue. Those aspects that within an air quality management system are 'purely' local in nature proved to be few. Thus, much of the subfunctions in the prime responsibilities of an air quality management system (e.g. public transport system, traffic management issues, police, monitoring systems) were regarded as exclusively metropolitan or, in any case, concurrent. From a two-tier perspective, this means that an upper tier of government should in principle be the ideal body to take charge of most air pollution functions. In defining the responsibilities of the GLA, the recent government's White Paper 'A Mayor and Assembly for London' resembles (under conventional assumptions) a 'top-heavy' metro scheme for air quality matters. There are very few sub-functions, such as commuter railways or parking regulations, over which the GLA will have little influence. Interestingly, the system that existed of concurrent powers shared by the former GLC and the boroughs regarding air quality monitoring and measurement functions (GLC 1983), will be re-installed as soon as the GLA starts operating.

Third, the empirical work clearly showed that air pollution in London needs to be tackled within a comprehensive or integrated fashion. According to conventional assumptions, the concept of metropolitan-local division recognised by metro systems can make an integrated functional management response to diverse services possible. In principle, the concentration of a whole function (or most of it) in one unit simplifies coordination and reduces boundary friction. Indeed, one of the main concerns that was found in London was not only the need for an integrated response, but the lack of coordination among boroughs, and among boroughs with other governmental and area-wide units. Not surprisingly, the main government reasons outlined for the creation of the GLA were about the need to provide a coordinated and 'strategic leadership' unit where issues such as transport and air quality could be tackled at a London-wide level (see DETR 1997b).

Fourth, the gap that was found in terms of coordination and standardisation of many air quality management aspects largely explains the emergence of diverse and alternative city-wide units across London. This, in turn, explains the existence of a complex set of institutions and actors (drawn from, but also beyond, government as envisaged by the local governance perspective) and the liaison jungle in which all of them operate. Although in the past local authorities in London have provided local services with other agencies, the range of responsibilities that they have held made them dominant within the overall system. This situation has changed with the lessening of the role of local authorities in London within the governance structure. As assumed by orthodox views, without a city-wide coordinating authority at the metropolitan level, the responsibility of some air quality London-wide functions has inevitably passed to central government and to other non-governmental area-wide bodies.

Finally, the results of the empirical investigation in London showed that some of the dilemmas or critical issues that are embedded within the local governance perspective as outlined by Stoker (1996b), are present in the current system of local government in London when dealing with air pollution. The main dilemmas include not only a clear need for a more democratic, accountable and legitimate system for managing air quality, but one in which local authorities may provide strategic leadership to coordinate local networks (see also Stewart & Stoker 1995b: 195; Newman & Thornley 1997: 984). It is important to note that the reasons that the British government has given for creating the GLA, in many ways resemble (albeit some differences exist) those that two-tier advocates have partly used for justifying the existence of metro schemes. Such reasons include: the need to fill the democratic deficit created by the abolition of the GLC in 1985, to provide strong strategic leadership, and to restore accountability (see DETR 1997b and 1998). Interestingly, these three particular aspects are regarded within the local governance perspective as critical issues which need to be addressed and solved. Thus, while an effective air quality management strategy in London should imply a more democratic accountability system (where power-holders are to be seen legitimate), there is a fundamental need to re-consider traditional orthodox assumptions in order to reduce the existing gap or divorce between the system of governance and the normative codes used to explain and justify government. Eventually, this may lead to a redefinition of such concepts as democracy, accountability and legitimacy.

Moving on to Mexico City, the discussion on reforming the structures of local government in this urban centre has not included air quality concerns (see Chapter 4). Nevertheless, in the light of the models of local government, the analysis of the results of how an area-wide coordinating authority in Mexico City has managed air quality provide some interesting findings when they are compared with the case of London. There are three main aspects regarding Mexico City's response to air pollution that need to be highlighted. First, Mexico City's system of local government is not as politically fragmented as London and operates with a lower degree of public-private partnerships and private networks (i.e. the governance perspective is less applicable). This is because air pollution in Mexico City has basically been a policy issue subjected to considerable governmental intervention, either from the central government or the DDF (now the GDF). Although the role of democratically elected local authorities may be strengthened within the governance structure, strong central government intervention is likely to remain in both urban centres. Second, although the findings in the interviews showed that in London and Mexico City there is strong central government intervention on air pollution matters, the underlying reasons for this are different. Whereas central government intervention on air quality issues in London has grown, largely due to political fragmentation at the local government level, in Mexico City such intervention has been due to the fact that air pollution was traditionally, until 1988, a federal function by constitutional mandate. Finally, while there is a fundamental need for a more democratic, accountable and legitimate system of local government in both urban centres, the underlying reasons for this are also different. While in London this mainly stems from an increased number of non-democratic area-wide agencies and public-private partnerships, in Mexico City it derives from the fact that the capital has been long governed by a non-democratic, area-wide and mainly centrally created government unit.

The outcome of how Mexico City's local government operates in relation to air quality management brings into light some interesting issues that relate to the traditional orthodox, public choice and local governance perspectives. First, orthodox views suggest that while the issues of equity and financial equalisation may be operated by both the central government and a metro authority, the latter has the advantage of also having greater local knowledge and accountability. In the case of Mexico City, the first part of this assumption proved to be right: the metropolitan problems of equity and financial equalisation were not issues of concern as most of the aspects which have broad catchment areas for the air quality management system were either under the responsibility of the central government or the GDF. However, the second part did not apply accordingly. The fact that the central government set up, for example, an effective, comprehensive and standardised air quality network system, suggests that the creation of a metro scheme cannot always be justified simply because local authorities have greater knowledge. Even more, in order to favour metro schemes over central government on accountability grounds, local authorities need to be democratically elected - something that did not happen in Mexico City until July 1997. Thus, such orthodox assumptions can only be relevant for creating a metro scheme in those urban centres where local authorities have traditionally been responsible for diverse air quality management sub-functions and can actually be accountable to the local electorate.

Second, while the findings corroborated the fact that air pollution is an area-wide issue, they did not show which aspects within an air quality management system should be functionally allocated to the delegated units in Mexico City. The only two issues where the participation of lower-tier authorities became evident were the issues of disseminating air quality information in their own local areas and of acting as organisations for channelising citizen demands to other levels of government. As suggested by public choice theorists, the monopolistic position that is likely to exist with a single governmental unit in a metropolitan area was present in the case of the former DDF. This was clearly manifested with the 1990 air quality management strategy where only very few (non-democratic and centrally appointed) local government individuals participated with other central and EdoMex authorities. Other overtly monopolistic behaviour by the former DDF relates to the issues of access and distribution of certain types of information - particularly regarding morbidity and mortality rates within the MZMC. On the one hand, a lack of motivation and weak political will clearly existed on the part of the DDF authorities (including also the EdoMex) to be more responsive, to carry out or stimulate research on air pollution effects on human beings. On the other hand, ample fragmentation of authority and overlapping jurisdictions where competition may be engendered to constrain the monopolistic behaviour of local authorities (as argued by public choice theorists), do not seem to ensure a solution to the problems of access and distribution of health information. Indeed, the empirical evidence showed that although still strictly confined to the world of academia, there are several universities and medical institutions that are carrying out that kind of research. Public access to and distribution of health information by metro authorities may be significantly a matter of pure political convenience, which has more to do with political competition (i.e. electoral matters) or responsibility avoidance, than with competitive and rival practices as envisaged by public choice advocates.

Third, while the need to create proper lower-tier authorities in the DF was evident for democratic and government-people proximity purposes, the existence of an area-wide authority operating in the whole DF proved to be quite convenient in terms of coordination for implementing diverse air quality measures. An integrated and coordinated government response to air pollution across an urban centre which requires an even participation of all local units, could be delayed or opposed under a system of jurisdictional fragmentation where local authorities may act according to their own political priorities and interests. Such was the case with one of the many municipalities of the EdoMex surrounding the DF, which showed resistance with regard to the implementation of 'A Day Without a Car' plan for the entire MZMC.

Fourth, the set up of a non-governmental area-wide unit - the CMPCCAVM - for the whole of the MZMC resembles in many ways what the orthodox tradition envisages as one of the many approaches to confront political fragmentation through cooperation among local governments: the metropolitan council (see Chapter 3). Although the CMPCCAVM did not arise only from local institutions, this area-wide mechanism has been constituted as a forum for deliberation and discussion as well as an advising and coordinating organisation for managing air quality and other environmental issues in the MZMC. The findings in the empirical work however, showed that consensus and concerted action among all participants was sometimes difficult to achieve. This situation seems to suggest that under a politically fragmented system as envisaged by public choice theorists, bargained or cooperative agreements for tackling air pollution would be even more problematic. It also suggests that if an existing metropolitan authority such as the former DDF is not large enough to contain the metropolitan system (i.e. an area of one or more urban centres affected by bad air quality) an organisation like the CMPCCAVM facilitates a quick governmental response to an emerging urban environmental problem - especially if no major local government reorganisation in a given metropolitan area is desired. Despite the flaws that are likely to appear with the creation of these type of organisations, the CMPCCAVM has the enormous advantage of bringing together metropolitan as well as state and central government bodies into one forum of discussion. The analysis of the results, thus, suggests that the best way to manage and coordinate air quality in a large metropolitan area is to reduce the number of government representatives. Ideally, though, these representatives must be democratically elected, and should allow other organisations to participate in air quality issues, such as affected surrounding local units, NGOs, or the private sector, in order to make the decision-making process as inclusive as possible.

Finally, while Mexico City's local authorities seemed to be operating in a complex and growing system of local governance, its system of local government (as already seen) is clearly a weak example of the dilemmas that are present within the local governance perspective when compared to the case of London. Indeed, the empirical work showed that the delivery and control of contracted out services in Mexico City, as well as strategic decision-making, were still very much within the domain of governmental structures (central and/or local), and not of private or voluntary networks. This does not imply that there were not private actors delivering public services. Such services as taxis or mini-buses are some of the areas of the system of public transport that have been subjected to some kind of contracting out. Likewise, all vehicle emission tests are carried out by private operators that have been given permission (or rather a concession) to provide such a service. While some public-private partnerships are likely to emerge in the near future (particularly after the 1997 local elections), the local government was still acting as the main actor responsible for city-management issues and public service provision. Moreover, the former DDF certainly occupied a relatively sovereign position and used strategic leadership in relation to other key policy players, despite existing tensions and difficulties with other central government units or institutions of civil society.

Summing up, although there is a need to move beyond the orthodox territory so as to widen the understanding on how local authorities operate in relation to air quality management, traditional assumptions (particularly the metro scheme approach) still provide a viable set of arguments about how to reorganise local government organisational structures in order to improve air quality (see Self 1997: 19; Stoker 1996a: 20). By contrast, the application of the public choice model to air quality management in terms of organisational structure is partly limited because this model takes as its starting point the diversity of individual preferences and the diverse nature of goods and services rather than the organisational structure (see Bish & Ostrom 1973: 17). However, as long as public choice assumptions discuss orthodox views, they provide a valuable approach for analysing how local government works regarding air quality management. While the existing range of policy responses to air pollution reinforces the 'hierarchical' approach, the current configuration of air quality as a public good emphasizes the externalities (negative and positive) and transaction costs that are associated with the public choice model which in turn leads one to consider the need for a metropolitan-type agency. Finally, the local governance approach provides an organisational framework for understanding the complex and growing system of 'networks' in which local governments may be operating. The latter may assist in considering diverse alternatives for managing air quality through, for example, community groups or local networks (e.g. cycling groups) promoting behaviourial change. In suggesting a new code for challenging past hierarchical and conventional modes of thinking, the local governance perspective assists in distinguishing diverse issues that local authorities need to take into account in their dealings with air pollution control.

Further avenues for research

This section seeks to identify those important and related areas of analysis which were not covered in this research due to time and space limitations. It mentions the modest contribution that a comparative study like this brings into local government studies and air quality policy research as well as commenting on the most relevant aspects that were experienced during the fieldwork process in London and Mexico City.

The conclusions of this study confirmed four major existing gaps in the academic literature on local government studies and air quality research that were identified in Chapter 1. First, although there are some comparative exercises on local government organisational structures (especially among Western societies), these are scarce within a North-South context. Second, there is very limited research in London and Mexico City (including comparative studies) on revising diverse proposals or models for local government reform in the light of the emerging urban environmental demands of the 1990s. Third, although diverse aspects of air quality management systems have been increasingly analysed, there have been very few comparative exercises in this regard. Finally, there is still limited research on how diverse systems of local government operate in order to improve air quality and bring about urban sustainability. This book makes a contribution towards closing these gaps but more research in each area still remains to be done.

The aftermath of doing a comparative study in London and Mexico City showed that while such an exercise enhances the understanding of the areas of analysis - e.g. local government systems and air quality management strategies - it also diminishes the possibility of making a more detailed examination of each case-study. This is partly explained due to the existing formal constraints of time and financing when carrying out a study like this, and partly because comparative exercises commonly demand paying less attention to specific technicalities but scanning for more general and comparative accounts. Clearly, London and Mexico City's local organisational structures and air quality management systems would benefit from further analysis in order to review more fully how local government in these two cities operates in relation to air pollution, particularly under the emerging systems of local governance.

Although the empirical data suggested the need for a variation of a metro scheme for dealing with air pollution (see Chapters 5-8), it does not prove definitively that with an area-wide authority (and lower-tier units) the problems of coordination, strategic leadership and representation, democracy and accountability, and public participation will be solved. Furthermore, some problems that the metro model will invariably have to face for managing air quality include functional allocation across different tiers of government (central, upper-tier and lower-tier), the metropolitan area to be covered, access to financial resources, and political strength of the metro authority vis à vis central government. It did show, however, that the best possible scenario across the whole local government organisational spectrum for better dealing with air pollution is the creation and adoption of some type of the metro scheme. This implies that in order to examine even further the implications of how diverse local organisational structures influence the capacity to address and improve the air pollution situation, there is a need to create a framework (not included in this research) for assessing government efficiency under diverse types of governmental arrangements. Thus, an area that also needs to be addressed in future studies includes setting up performance indicators to measure efficiency and effectiveness according to each local authority's own air quality management objectives.

From a theoretical point of view, although there are many issues in relation to air quality management that were raised in the book (particularly regarding public transportation systems and traffic management) they were not all addressed within the theories explained in Chapter 3 due to space limitations. Further research on the analysis of such theories and air pollution needs to take into account air quality management aspects raised in this study in order to examine the weaknesses of all approaches, as well as to take the best from each of them for best governmental practice.

The analysis of the outcome of this research also showed that despite the similarities and differences that may exist between two or more distinct urban centres across the world, it is possible to carry out a pragmatic, empirically-based and comparative study. While this book acknowledges some methodological limitations and weaknesses due to the need for a more detailed description of each case-study and a wider explanation of diverse models of local government, it also provides an enhanced understanding of some of the steps that are required for doing comparative studies. Certainly, one of the most important aspects of the latter is the need to carry out empirical fieldwork in the urban areas that are subjected to analysis. In so doing, one must be prepared to experience a number of unexpected events and to adapt to diverse procedures both for collecting data from governmental and non-governmental bodies, and carrying out structured and semi-structured interviews. So, for example, in the case of London, all the semi-structured interviews were carried out as scheduled and on the arranged date and time. By contrast, in Mexico City, while carried out as scheduled, at least half of the semi-structured interviews with central and local authorities were either postponed for a following day or usually never on time (i.e. would take place one or even two hours later). It is also of utmost importance to take into account that both updating interviews, and data collection, may be fundamental during the whole research process.

Finally, a question that was not addressed in this present study is the examination of social consciousness and/or public participation in air quality aspects precisely because many of the air pollution problems are caused by motorists. This would inevitably involve the analysis of diverse mechanisms for increasing environmental education; for example, through disseminating information containing the dangers that air pollution may cause to human beings, flora and fauna, and the built environment. London and Mexico City inhabitants require environmental awareness on air quality management issues so they can make proper use of their motor vehicles and/or switch to other means of public transport. However, it is the ignorance and apathy of some government authorities (at central and local levels) which is in more urgent need of being addressed.

Conclusion

By the year 2000, London and Mexico City will continue to suffer from perceived poor air quality. While in the near future the levels of air pollution concentration may not increase from current levels, it is quite unlikely that any significant improvement will take place in any of these two urban centres. Although a few pollution episodes may be experienced in London in the coming years (particularly during the wintertime) it is far from suffering the high and health-risk levels of pollution that exist in Mexico City. In spite of available (though still limited) information on present morbidity and mortality rates in these two cities, the UK and Mexican governments have not changed their policies toward road building expansion, which has in turn favoured the use of private vehicles over public transport systems. London and Mexico City car users have not significantly shifted as yet to other means of transport so as to make a positive impact on bringing down air pollution levels. Seemingly, this situation is not expected to change in the short time.

The process of local government reorganisation that is currently taking place in London and Mexico City is leading local authorities in these two cities to further their participation in diverse air quality management issues. The adoption of metro schemes in both urban centres does not guarantee achieving and maintaining healthy urban air quality in the short term, but it is certainly a step in the right direction for making the governmental response more adequate in terms of coordination, strategic leadership and representation, and public participation (i.e. democracy and accountability). The degree to which these metro bodies may or may not succeed will depend a great deal on the figure of the mayor who will have to dominate political life in the capital as well as to administrate or manage air quality, jointly with other local and central authorities, NGOs, the private sector, and the public. The changes in the organisational structures of local government in London and Mexico City are encouraging and optimistic for better dealing with air pollution; it is now up to politicians and citizens to make institutional and policy outcomes work and thus improve air quality in both urban centres.

Bibliography

Akin Aina, T. et al (eds) (1991), 'Rethinking Local Government – Views from the Third World', *Environment and Urbanization,* vol. 3, No. 1, April, pp. 3-7.

Alderfer, H.F. (1964), *Local Government in Developing Countries,* McGraw-Hill, New York.

Anderson, W. (1925), *American City Government,* American Political Science Series, Pitman, London.

Anderson, W. (1934), *The Units of Government in the United States,* No. 42, Public Administration Service, Chicago.

Anderson, W. and Weidner, E.W. (1950), *American City Government,* Revised Edition, Henry Holt, New York.

Ashby, E. and Anderson, M. (1981), *The Politics of Clean Air,* Clarendon, Oxford.

Association of London Government (ALG) (1997), *A Breath of Fresh Air: Air Quality in London,* July, Association of London Government, London.

Atkinson, A. (1991), 'Environment and development: concepts and practices in transition', *Public Administration and Development,* vol. 11, pp. 401-413.

Ball, S. and Bell, S. (1994), *Environmental Law,* Second Edition, Blackstone, London.

Ball, S. and Bell, S. (1995), *Environmental Law,* Third Edition, Blackstone, London.

Barlow, I.M. (1981), *Spatial Dimensions of Urban Government,* Research Studies Press, Wiley, Chichester.

Barlow, I.M. (1991), *Metropolitan Government,* Routledge, London.

Barrera Echeverri, G. et al (1993), 'Sistema Operativo de Contingencia Ambiental del Sistema Nacional de Salud', in Secretaría de Salubridad y Asistencia and Departamento del Distrito Federal, *Acciones Desarrolladas para Evaluar el Efecto en la Salud por la Contaminación Atmosférica en el Valle de México,* Sistema Nacional de Salud, Mexico City, pp. 39-51.

Barry, J. (1996), 'Sustainability, Political Judgement and Citizenship: connecting green politics and democracy', in B. Doherty and M. De Geus (eds), *Democracy and Green Political Thought,* Routledge, London, pp. 115-131.

Bartone, C. et al (1994), *Toward environmental strategies for cities: policy considerations for urban environmental management in developing countries,* Urban Management Programme, United Nations Development Programme, United Nations Conference on Human Settlements, World Bank, 18, World Bank, Washington D.C.

Bauer, M. and Quintanilla, J. (1995), 'Emissions of the energy chains in the Mexican energy system', Paper presented at *Seminario Internacional sobre Sustentabilidad Urbana y el Manejo de la Cuenca Atmosférica en la Zona*

Metropolitana del Distrito Federal y del Estado de México, May 22nd-25th, Mexico City.

Bell, S. (1993), *Capital Killer II: still fuming over London's traffic pollution,* London Boroughs Association (LBA), London.

Ben-Elia, N. (ed) (1996a), *Strategic Changes and Organisational Reorientations in Local Government,* Macmillan, Basingstoke and London.

Ben-Elia, N. (1996b), 'Introduction', in N. Ben-Elia (ed), *Strategic Changes and Organisational Reorientations in Local Government,* Macmillan, Basingstoke and London, pp. 1-20.

Bennet, G. (ed) (1991), *Air Pollution Control in the European Community,* International Environmental Law and Policy Series, Graham and Trotman, London.

Bennet, R.J. (ed) (1993a), *Local Government in the New Europe,* Belhaven, London and New York.

Bennet, R.J. (1993b), 'Local government in Europe: common directions of change', in R.J. Bennet (ed), *Local Government in the New Europe,* Belhaven, London and New York.

Birnie, P.W. and Boyle, A.E. (1992), *International Law and the Environment,* Clarendon, Oxford.

Bish, R.L. (1971), *The Public Economy of Metropolitan Areas,* Markham, Chicago.

Bish, R.L. and Ostrom, V. (1973), *Understanding Urban Government,* American Enterprise Institute for Public Policy Research, Washington D.C.

Blowers, A. (ed) (1993a), *Planning for a Sustainable Environment,* Town and Country Planning Association, Earthscan, London.

Blowers, A. (1993b), 'The Time for Change', in A. Blowers (ed), *Planning for a Sustainable Environment,* Town and Country Planning Association, Earthscan, London, pp. 1-18.

Blowers, A. (1993c), 'Pollution and Waste – A Sustainable Burden?', in A. Blowers (ed), *Planning for a Sustainable Environment,* Town and Country Planning Association, Earthscan, London, pp. 69-92.

Bollens, J.C. and Schmandt, H.J. (1965), *The Metropolis: its people, politics and economic life,* Harper and Row, New York.

Borja, J. (1992), 'Past, Present, and Future of Local Government in Latin America', in R.M. Morse and J.E. Hardoy (eds), *Rethinking the Latin American City,* Woodrow Wilson Center, Washington D.C., pp. 131-143.

Brañes, R. (1994), *Manual de Derecho Ambiental Mexicano,* Fundación Mexicana para la Educación Ambiental, Fondo de Cultura Económica (FCE), Mexico City.

Bravo Alvarez, H. (1987), *La Contaminación del Aire en México,* Universo Veintiuno, Mexico City.

Bravo Alvarez, H. (1993), 'El Transporte Contra la Contaminación Atmosférica: Presente y Futuro al Año 2000 en la Zona Metropolitana de la Ciudad de

México', in Greenpeace Mexico, *El Transporte y la Contaminación,* Greenpeace Mexico, Mexico City, pp. 7-9.

Bravo Alvarez, H. and Torres Jardón, R. (1995), 'Prioridades en Contaminación Atmosférica en la Ciudad de México: ozono y partículas respirables', in *Foro de Consulta Popular Medio Ambiente y Desarrollo Sustentable en el Valle de México,* Mexico City.

Breheny, M.J. (1992a), 'Sustainable Development and Urban Form: An Introduction', in M.J. Breheny (ed), *Sustainable Development and Urban Form,* Pion, London, pp. 1-23.

Breheny, M.J. (1992b), 'Towards Sustainable Urban Development', in A.M. Mannion and S.R. Bowlby (eds), *Environmental Issues in the 1990s,* Wiley, Chichester, pp. 277-290.

Breheny, M.J. and Rookwood, R. (1993), 'Planning the Sustainable City Region', in A. Blowers (ed) *Planning for a Sustainable Environment,* Town and Country Planning Association, Earthscan, London, pp. 150-189.

Bridges, O. (1995), 'Comparison of air quality in the UK and Russia', *Environmentalist,* 15 (2), pp. 139-146.

Brimblecombe, P. (1987), *The Big Smoke,* Routledge, London.

Brown, T. (1993), 'The Future of Air Quality Monitoring Networks', *Local Air Quality Management,* Winter, National Society for Clean Air (NSCA), London, pp. 3-4.

Bryman, A. and Burgess, R.G. (1994), 'Development in qualitative data analysis: an introduction', in A.Bryman and R.G. Burgess (eds), *Analysing Qualitative Data,* Routledge, London and New York, pp. 1-17.

Buchanan, J.M. and Tullock, G. (1962), *The Calculus of Consent,* Ann Arbor, University of Michigan.

Burgoa, I. (1985), *Derecho Constitucional Mexicano,* Sixth Edition, Porrúa, Mexico City.

Button, J. (1988), *A dictionary of green ideas: vocabulary for a sane and sustainable future,* Routledge, London.

Cabannes, Y. (1997), 'Interview with Mexico City's Governor-Elect Cuauhtémoc Cárdenas', *The Urban Age,* vol. 5, no. 2, The World Bank, Washington D.C., pp. 26.

Calderón-Garcidueñas, L. et al (1992), 'Histopathologic Changes of the Nasal Mucosa in Southwest Metropolitan Mexico City Inhabitants', *American Journal of Pathology,* vol. 140, no. 1, January, pp. 225-232.

Calvillo, A. (1993), 'Seminario el Transporte y la Contaminación en la Ciudad de México', in Greenpeace Mexico, *El Transporte y la Contaminación,* Greenpeace Mexico, Mexico City, pp. 3-6.

Calvillo, A. (1995), 'Propuesta Alternativa de Transporte Público y Vialidad para la Ciudad de México', Paper presented at *Seminario Internacional sobre Sustentabilidad Urbana y el Manejo de la Cuenca Atmosférica en la Zona*

Metropolitana del Distrito Federal y del Estado de México, May 22nd-25th, Mexico City.

Campos Ruíz, H. et al (1993), 'Políticas para la Prevención y Control de la Contaminación: el caso de la Ciudad de México', in *Diagnóstico de los países centroamericanos en torno a la implementación de los problemas de la política ambiental,* ZöV Zentralstelle für öffentliche Verwaltung, Berlin, pp. 59-74.

Cancino Aguilar, M.A. (1994), 'El Marco Jurídico para la Prevención y Control de la Contaminación Atmosférica', in J.J. González Márquez (ed), *Derecho Ambiental,* Universidad Autónoma Metropolitana (UAM), Mexico City, pp. 105-159.

Carlson, E. (1996), 'The Legacy of Habitat II', *The Urban Age,* vol. 4, no. 2, August, World Bank, Washington D.C., pp. 1-6.

City of Westminster (1993), *The State of the Environment in Westminster,* Summary Report, December, City of Westminster, London.

Clarke, M. (1997), 'Local government and the new public management: A review', in H. Davis et al, *New perspectives in local governance,* Joseph Rowntree Foundation, York, pp. 37-73.

Clarke, M. and Stewart, J. (1990), *General Management in Local Government: Getting the Balance Right,* Longman, Harlow.

Clarke M., and Stewart, J. (1996), 'The Learning Local Authority: British Local Authorities in the Post-Thatcherite Era', in N. Ben-Elia (ed), *Strategic Changes and Organisational Reorientations in Local Government,* Macmillan, Basingstoke and London, pp. 41-56.

Cochrane, A. (1993), *Whatever Happened to Local Government?,* Open Univeristy, Buckingham.

Cohen, M. (1996), 'Habitat II: A Critical Assessment', *The Urban Age,* vol. 4, no. 2, August, World Bank, Washington D.C., pp. 6-21.

Collins, Ch.O. and Scott, S.L. (1993), 'Air Pollution in the Valley of Mexico', *Geographical Review,* April, vol. 83, no. 2, pp. 119-133.

Comisión Metropolitana para la Prevención y Control de la Contaminación Ambiental en el Valle de México (CMPCCAVM) (1992), *¿Qué estamos haciendo para combatir la contaminación del aire en el Valle de México?,* Comisión Metropolitana para la Prevención y Control de la Contaminación Ambiental en el Valle de México, Mexico City.

Comisión Metropolitana para la Prevención y Control de la Contaminación Ambiental en el Valle de México (CMPCCAVM) (1994a), *Avances a Junio de 1994 del Programa Integral Contra la Contaminación Atmosférica,* Comisión Metropolitana para la Prevención y Control de la Contaminación Ambiental en el Valle de México, Mexico City.

Comisión Metropolitana para la Prevención y Control de la Contaminación Ambiental en el Valle de México (CMPCCAVM) (1994b), *Air pollution in Mexico City. Present situation,* Comisión Metropolitana para la Prevención y Control de la Contaminación Ambiental en el Valle de México, Mexico City.

Comisión Metropolitana para la Prevención y Control de la Contaminación Ambiental en el Valle de México (CMPCCAVM) (1994c), *La Contaminación Atmosférica en el Valle de México,* Comisión Metropolitana para la Prevención y Control de la Contaminación Ambiental en el Valle de México, Mexico City.

Comisión Metropolitana para la Prevención y Control de la Contaminación Ambiental en el Valle de México (CMPCCAVM) (1995a), *Medidas de invierno 1994-1995: Informe final,* April, Comisión Metropolitana para la Prevención y Control de la Contaminación Ambiental en el Valle de México, Mexico City.

Comisión Metropolitana para la Prevención y Control de la Contaminación Ambiental en el Valle de México (CMPCCAVM) (1995b), *Informe Mensual de la Calidad del Aire,* May, Comisión Metropolitana para la Prevención y Control de la Contaminación Ambiental en el Valle de México, Mexico City.

Commission of the European Communities (CEE) (1990), *Green Paper on the Urban Environment,* Commission of the European Communities, Luxembourg.

Committee for Economic Development (CED) (1966), *Modernising Local Government,* July, Committee for Economic Development, New York.

Committee for Economic Development (CED) (1970), *Reshaping Government in Metropolitan Areas,* February, Committee for Economic Development, New York.

Conservative (1997), *The Conservative Manifesto 1997: You Can Only be Sure with the Conservatives,* Conservative Central Office, London.

Constitution (1997), *Constitución Política de los Estados Unidos Mexicanos,* Porrúa, Mexico City.

Cox, K.R. and Nartovicz, F.Z. (1980), 'Jurisdictional Fragmentation in the American Metropolis', *International Journal of Urban and Regional Research,* vol. IV, no. 2, Edward Arnold, London, pp. 196-211.

Cuchillo, M. and Morata, F. (1991), *Organización y funcionamiento de las áreas metropolitanas: un análisis comparado,* Ministerio para las Administraciones Públicas, Madrid.

Daly, H.E. (1993), 'The Perils of Free Trade', *Scientific American,* November, pp. 24-29.

Davey, K. (1996), 'The structure and functions or urban government' in K. Davey, *Urban Management,* Avebury, Aldershot, pp. 47-102.

Davis, D.E. (1994), *Urban Leviathan: Mexico City in the Twentieth Century,* Temple University, Philadelphia.

Davis, H. et al (1997), *New perspectives in local governance,* Joseph Rowntree Foundation, York.

De Geus, M. (1996), 'The Ecological Restructuring of the State', in B. Doherty and M. de Geus (eds), *Democracy and Green Political Thought,* Routledge, London, pp. 188-211.

Dearlove, J. (1979), *The Reorganisation of British Local Government,* Cambridge University, Cambridge.

Dente, B. (1990), 'Metropolitan Governance Reconsidered, or How to Avoid Errors of the Third Type', *Governance,* vol. 3, no. 1, January, pp. 55-74.

Departamento del Distrito Federal (DDF) (1992), *Cuadernos de la Reforma Política de la Ciudad de México: El Gobierno de la Ciudad de México 1524-1992,* Departamento del Distrito Federal, Mexico City.

Departamento del Distrito Federal (DDF) (1993), *Análisis de la propuesta del Gobierno del Presidente Carlos Salinas de Gortari para la Reforma Política del Distrito Federal,* April, Procuraduría Fiscal del Distrito Federal, Departamento del Distrito Federal, Mexico City.

Departamento del Distrito Federal (DDF) (1994), *Red Automática de Monitoreo Atmosférico,* December, Dirección de la Red Automática de Monitoreo Atmosférico (RAMA), Departamento del Distrito Federal, Mexico City.

Departamento del Distrito Federal (DDF) (1995), *Programa Integral de Transporte 1995-2000,* Departamento del Distrito Federal, Mexico City.

Departamento del Distrito Federal (DDF) (1996), *Programa para Mejorar la Calidad del Aire en el Valle de México 1995-2000,* March, Departamento del Distrito Federal, Mexico City.

Department of the Environment (DoE) (1983), *Steamlining the Cities,* Cmnd 9063, Her Majesty's Stationery Office (HMSO), London.

Department of the Environment (DoE) (1994), *Sustainable Development: The UK Strategy,* January, Department of the Environment, Her Majesty's Stationery Office (HMSO), London.

Department of the Environment (DoE) (1995), *Air Quality: Meeting the Challenge,* January, Department of the Environment, London.

Department of the Environment (DoE) (1996), *The United Kingdom National Air Quality Strategy – Consultation Draft,* August, Department of the Environment, London.

Department of the Environment (DoE) (1997), *The United Kingdom National Air Quality Strategy,* Cm 3587, March, Department of the Environment, The Stationery Office, London.

Department of the Environment, Transport and the Regions (DETR) (1997a), *Keeping Buses Moving,* Local Transport Note 1/97, Department of the Environment, Transport and the Regions, The Stationary Office, London.

Department of the Environment, Transport and the Regions (DETR) (1997b), *New Leadership for London,* Cm 3724, July, Department of the Environment, Transport and the Regions, The Stationary Office, London.

Department of the Environment, Transport and the Regions (DETR) (1998), *A Mayor and Assembly for London: The Government's proposals for modernising the governance of London,* Department of the Environment, Transport and the Regions, The Stationary Office, London.

Department of Health (DoH) (1991), *Ozone,* First Report, Advisory Group on the Medical Aspects of Air Pollution Episodes, Department of Health, Her Majesty's Stationery Office (HMSO), London.

Department of Health (DoH) (1992), *Sulphur Dioxide, Acid Aerosols and Particulates,* Second Report, Advisory Group on the Medical Aspects of Air Pollution Episodes, Department of Health, Her Majesty's Stationery Office (HMSO), London.

Department of Health (DoH) (1993), *Oxides of Nitrogen,* Third Report, Advisory Group on the Medical Aspects of Air Pollution Episodes, Department of Health, Her Majesty's Stationery Office (HMSO), London.

Department of Health (DoH) (1995), *Health Effects of Exposures to Mixtures of Air Pollutants,* Fourth Report, Advisory Group on the Medical Aspects of Air Pollution Episodes, Department of Health, Her Majesty's Stationery Office (HMSO), London.

Department of Health (DoH) (1997), *Handbook on Air Pollution and Health,* Committee on the Medical Effects of Air Pollutants, Department of Health, The Stationery Office, London.

Department of Transport (DoT) (1996), *A Transport Strategy for London,* Department of Transport and Government Office for London, Her Majesty's Stationery Office (HMSO), London.

Díaz Alfaro, S. (1992), 'El Distrito Federal Mexicano: Breve Historia Constitucional', in *Estudios Jurídicos en torno a la Constitución Mexicana de 1917 en su Septuagésimo Aniversario,* Instituto de Investigaciones Jurídicas and Universidad Nacional Autónoma de México (IIJ-UNAM), Mexico City, pp. 197-216.

Díaz Díaz, D. and Perló Cohen, M. (1994), *Retos y Propuestas: Ciudad de México,* Fundación Mexicana Cambio XXI Luis Donaldo Colosio, Mexico City.

Dobson, F. (1995?), *Creating a Bad Atmosphere: the Tory threat to Britain's air quality,* a report by Frank Dobson MP, Shadow Environment Secretary, Labour, London.

Doherty, B. and De Geus, M. (eds) (1996), *Democracy and Green Political Theory,* Routledge, London, pp. 1-15.

Dowding, K. (1996), 'Public Choice and Local Governance', in D. King and G. Stoker (eds), *Rethinking Local Democracy,* Government Beyond the Centre, Economic and Social Research Council (ESRC), Macmillan, Basingstoke, pp. 50-66.

Downs, A. (1967), *Inside Bureaucracy,* Little Brown, Boston.

Dunleavy, P. (1980), *Urban Political Analysis,* Macmillan, London and Basingstoke.

Dunleavy, P. (1991), *Democracy, Bureaucracy and Public Choice,* Harvester Wheatsheaf, London.

Dunleavy, P. and O'Leary, B. (1987), *Theories of the State,* Macmillan, Basingstoke.

Edralin, J.S. (1996), 'Local Governance and Local Economic Development Research Findings on Capacity-Building', in J.S. Edralin (ed), *Local Governance and Local Economic Development: Capacity-Building. Case Studies in Asia and Latin America,* United Nations Centre for Regional Development (UNCRD), Nagoya, pp. 3-20.

Ehrlich, P.R. (1969), *The Population Bomb,* Ballantine Books, New York.

Ehrlich, P.R. and Ehrlich, A.H. (1970), *Population, Resources, Environment: issues in human ecology,* Freeman, San Francisco.

Elkin, T. et al (1991), *Reviving the City: Towards Sustainable Urban Development,* Friends of the Earth, London.

Elsom, D. (1992), *Atmospheric Pollution: A Global Problem,* Second Edition, Blackwell, Oxford.

Elsom, D. (1996), *Smog Alert: Managing Urban Air Quality,* Earthscan, London.

Environment Act (EA) (1995), *Environment Act 1995,* Chapter 25, Her Majesty's Stationery Office (HMSO), London.

Environmental Protection Act (EPA) (1990), *Environmental Protection Act 1990,* Chapter 43, Her Majesty's Stationary Office (HMSO), London.

Eskeland, G.S. et al (1994), 'Energy Pricing and Air Pollution: Econometric Evidence from Manufacturing in Chile and Indonesia', *Working Papers,* 1323, July, World Bank, Washington D.C.

Espinoza, L. (Chamalú) (1993), *Ecología Chamánica,* Second Edition, Colección Nueva Consciencia, Obelisco, Barcelona.

Estatuto de Gobierno del Distrito Federal (EGDF) (1994), *Estatuto de Gobierno del Distrito Federal,* D.O. 26/VII/1994, Mexico City.

Ezcurra, E. (1990), 'The Basin of Mexico City', in B.L. Turner II (ed), *The Earth as Transformed by Human Action,* Cambridge University, Cambridge, pp. 577-588.

Ezcurra, E. (1995), 'Reunión sobre el Manejo de la Cuenca Atmosférica de la Ciudad de Mëxico: el marco histórico-urbano', Paper presented at *Seminario Internacional sobre Sustentabilidad Urbana y el Manejo de la Cuenca Atmosférica en la Zona Metropolitana del Distrito Federal y del Estado de México,* May 22nd-25th, Mexico City.

Finer, H. (1993), *English Local Government,* Methuen, London.

Flynn, N. et al (1985), *Abolition or Reform?,* Allen and Unwin, London.

Flynn, N. and Leach, S. (1984), *Joint boards and joint committees: an evaluation,* Institute of Local Government Studies, University of Birmingham, Birmingham.

Foster, C.D. et al (1980), *Local Government Finance in a Unitary State,* Allen and Unwin, London.

Fried, R.C. (1972), 'Mexico City' in W.A. Robson and D.E. Regan (eds), *Great Cities of the World,* vol. II, Allen and Unwin, London, pp. 647-687.

Fried, R.C. and Rabinovitz, R. (1980), *Comparative Urban Politics,* Contemporary Comparative Politics Series, Prentice-Hall, New Jersey.

Friends of the Earth (FoE) (1991), *Report: Air Quality and Health,* July, Friends of the Earth, London.

Friends of the Earth (FoE) (1994), *Transport-Related Atmospheric Pollution in London: Summary of Evidence from Friends of the Earth,* Friends of the Earth, London.

Gamboa de Buen, J. (1994), *Ciudad de México, una visión,* Fondo de Cultura Económica (FCE), Mexico City.

Gandy, M. (1993), *Recycling and Waste: An exploration of contemporary environmental policy,* Avebury, Aldershot.

García Guadilla, M.P. (ed) (1991), *Ambiente, Estado y Sociedad,* Universidad Simón Bolívar, Centro de Estudios del Desarrollo (CENDES), Caracas.

García, R. et al (1998), 'Aldehídos: contaminantes atmosféricos', *Teorema,* year 4, no. 16, March-May, Mexico City, pp. 30-31.

Gibbon, G. and Bell, R.W. (1939), *History of the L.C.C. 1889-1939,* Macmillan, London.

Gilbert, R. et al (1996), *Making Cities Work,* Earthscan, London.

Goldsmith, E. et al (1972), *A Blueprint for Survival,* Penguin, London.

Golembiewski, R.T. (1977), 'A Critique of 'Democratic Administration' and its supporting Ideation', *American Political Science Review,* vol. LXXI, no. 4, December, pp. 1488-1507 and 1526-1531.

González Márquez, J.J. and Cancino Aguilar, M.A. (1994), 'La Distribución de Competencias en Materia Ambiental', in J.J. González Márquez (ed), *Derecho Ambiental,* Universidad Autónoma Metropolitana (UAM), Mexico City, pp. 25-65.

Goodwin, M. and Painter, J. (1996), 'Local governance, the crises of Fordism and the changing geographies of regulation', *Transactions, Institute of British Geographers,* New Series, vol. 21, no. 4, Dorchester, pp. 635-648.

Gordon, J. (1993), *Uncorking the Genie: Local Government and UNCED follow up,* Global Environment Research Centre, London, pp. 3-23.

Gossop, Ch. and Webb, A. (1993), 'Getting Around: Public and Private Transport', in A. Blowers (ed), *Planning for a Sustainable Environment,* Town and Country Planning Association, Earthscan, London.

Gottman, J. (1995), 'Introduction: Why Metropolitan Organisation?', in L.J. Sharpe (ed), *The Government of World Cities: the future of the metro model,* Wiley: Chichester, pp. 1-9.

Greater London Council (GLC) (1983), *Thirty Years On: A Review of Air Pollution in London,* Greater London Council, London.

Greater London Council (GLC) (1984), *Responses by GLC Committees to the Government's White Paper 'Steamlining the Cities',* Greater London Council, London.

Greenpeace (1993), *Ciudades Libres de Carros,* Greenpeace International, Vienna.

Greenwood, J. and Wilson, D. (1989), *Public Administration in Britain Today,* Second Edition, Unwin Hyman, London.

Greig, A.D. (1993), 'Urban Sustainability: An annotated bibliography', *Research Papers,* Department of Geography, London School of Economics, London.

Grubb, M. et al (1993), *The Earth Summit Agreements: A Guide and Assessment,* The Royal Institute of International Affairs, Earthscan, London.

Gunlicks, A.B. (ed) (1981), *Local Government Reform and Reorganisation,* National University, Kennikat, Port Washington and London.

Hall, P. (1984), *The World Cities,* Third Edition, Weindenfeld and Nicolson, London.

Hall, P. (1989), *London 2001,* Unwin Hyman, London.

Hallman, H.W. (1977), *Small and Large Together: Governing the Metropolis,* Sage, Beverly Hills and London.

Hardie, R.W. et al (1995), 'Development of a methodology for evaluating air pollution options for improving the air quality in Mexico City', *The Science of the Total Environment,* vol. 1, pp. 1-7?

Hardoy, J.E. et al (1992), *Environmental problems in Third World cities,* Earthscan, London.

Harré, R. (1979), *Social Being,* Blackwell, Oxford.

Harré, R. (1993), *Social Being,* Second Edition, Blackwell, Oxford.

Hatcher, R.L. (1996), 'Local Indicators of Sustainability: Measuring the Human Ecosystem', in B. Nath et al (eds), *Sustainable Development,* VUBPRESS, Brussels, pp. 207-232.

Haughton, G. and Hunter, C. (1994), *Sustainable Cities,* Regional Studies Association, Kingsley, London.

Hebbert, M. (1991), 'The Borough Effect in London's Geography' in K. Hoggart and D.R. Green (eds), *London: a new metropolitan geography,* Edward Arnold, London, pp. 191-206.

Hebbert, M. (1992), 'Governing the Capital', in A. Thornley (ed), *The Crisis of London,* Routledge, London, pp. 134-148.

Hebbert, M. and Travers, T. (1988), *The London Government Handbook,* Greater London Group, Casell, London.

Her Majesty's Stationery Office (HMSO) (1996), 'Britain's Legal Systems', *Aspects of Britain,* Publishing Services, Central Office of Information, Her Majesty's Stationery Office, London.

Hernández Avila, M. (1995?), *Producción Bibliográfica en el Area de Salud Ambiental,* Centro de Investigaciones de Salud Pública, Cuernavaca.

Hesse, J.J. and Sharpe L.J. (1990/91), 'Local Government in International Perspective: Some Comparative Observations', in J.J. Hesse (ed), *Local Government and Urban Affairs in International Perspective,* Nomos Verlagsgesellschaft, Baden-Baden, pp. 603-621.

Hicks, U.K. (1974), *The Large City: A World Problem,* Macmillan, London and Basingstoke.

Hirsch, W.Z. (1964), 'Local vs. Area-wide urban Government Services', *National Tax Journal,* vol. XVII, no. 4, December, Tax Institute of America, Chicago, pp. 331-339.

Hoggart, K. And Green, D.R. (1991), *London: a new metropolitan geography,* Edward Arnold, London.

Hood, C. (1991), 'A Public Management for All Seasons', *Public Administration,* vol. 69, Spring, pp. 3-19.

Humes IV, S. (1991), *Local Governance and National Power*, International Union of Local Authorities (IULA) and Harvester Wheatsheaf, Hemel Hempstead.

Humes, S. and Martin, E. (1969), *The Structure of Local Government,* International Union of Local Authorities (IULA), The Hague.

Hundsalz, M. (1996), 'The Habitat Debate: Goals, Commitments and Action Plans', *Habitat Debate,* September/December, vol. 2, no. 3 / 4, United Nations Conference for Human Settlements (UNCHS) (Habitat), Nairobi, pp. 6-8.

Icazuriaga Montes, C. (1992), *La Metropolización de la Ciudad de México a través de la instalación industrial,* CIESAS, Mexico City.

Instituto Nacional de Estadística, Geografía e Informática (INEGI) (1992), 'Ciudad de México (Area Metropolitana). Resultados Definitivos. Tabulados Básicos', in *XI Censo General de Población y Vivienda 1990,* Instituto Nacional de Estadística, Geografía e Informática, Aguascalientes.

Instituto Nacional de Estadística, Geografía e Informática (INEGI) (1994a), *Estadísticas del Medio Ambiente,* Instituto Nacional de Estadística, Geografía e Informática, Aguascalientes.

Instituto Nacional de Estadística, Geografía e Informática (INEGI) (1994b), *Estadísticas Históricas de México,* Third Edition, vol. I and II, Instituto Nacional de Estadística, Geografía e Informática, Aguascalientes.

Jackson, W.E. (1965), *Achievement: A Short History of the London County Council,* Longman, London.

John, P. (1990), *Recent Trends in Central-Local Government Relations,* Policy Studies Institute for Joseph Rowntree Foundation, London.

Johnston, R.J. and Pattie, C.J. (1996), 'Local Government in Local Governance: the 1994-1995 Restructuring of Local Government in England', *International Journal of Urban and Regional Research,* vol. 20, no.4, Blackwell, Oxford, pp. 671-696.

Jones, V. (1942), *Metropolitan Government,* University of Chicago, Chicago.

Jones, V. and O'Donnell, P., 'Metropolitan Areas', in D.C. Rowat (ed), *International Handbook on Local Government Reorganisation,* Aldwych, London, pp. 539-553.

Kandell, J. (1988), *La capital: the biography of Mexico City,* Random House, New York.

Keating, M. (1993), *The Earth Summit's Agenda for Change,* Centre for Our Common Future, Geneva.

Keating, M. (1995), 'Size, Efficiency and Democracy: Consolidation, Fragmentation and Public Choice', in D. Judge et al (eds), *Theories of Urban Politics,* Sage, London, pp. 117-134.

King, D. (1995), 'From the Urban Left to the New Right: Normative Theory and Local Government', in J. Stewart and G. Stoker (eds), *Local Government in the 1990s,* Government Beyond the Centre, Macmillan, London, pp. 228-248.

Kooiman, J. (ed) (1993), *Modern Governance: New Government-Society Interactions,* Sage, London.

Labour (1997), *The Labour Manifesto 1997: new Labour, because Britain deserves better,* Labour Party, London.

Leach, S. (ed) (1992), *Strengthening local government in the 1990s,* Longman, Harlow.

Leach, S. et al (1994), *The Changing Organisation and Management of Local Government,* Government Beyond the Centre, Macmillan, Basingstoke and London.

Leemans, A.F. (1970), *Changing Patterns of Local Government,* International Union of Local Authorities (IULA), The Hague.

Leff, E. (1991), 'Análisis Sociológico del Movimiento Ambientalista en América Latino', in M.P. García Guadilla (ed), *Ambiente, Estado y Sociedad,* Universidad Simón Bolívar, Centro de Estudios del Desarrollo (CENDES), Caracas, pp. 133-150.

Leff, E. (1995), *Green Production: Toward an Environmental Rationality,* Guildford, New York.

Levinson, A. and Shetty, S. (1992), 'Efficient Environmental Regulation: Case Studies of Urban Air Pollution', *Working Papers,* August, World Bank, Washington D.C.

Ley Ambiental del Distrito Federal (LADF) (1996), *Ley Ambiental del Distrito Federal,* G.O. 8/VII/1996, Mexico City.

Ley Federal de Protección al Ambiente (LFPA) (1982), *Ley Federal de Protección al Ambiente,* D.O. 27/I/1982, Mexico City.

Ley General del Equilibrio Ecológico y la Protección al Ambiente (LGEEPA) (1988), *Ley General del Equilibrio Ecológico y la Protección al Ambiente,* D.O. 28/I/1988, Mexico City.

Ley Orgánica de la Administración Pública del Distrito Federal (LOAPDF) (1994), *Ley Orgánica de la Administración Pública del Distrito Federal,* G.O. 31/XII/1994, Mexico City.

Ley Orgánica del Departamento del Distrito Federal (LODDF) (1978), *Ley Orgánica del Departamento del Distrito Federal,* D.O. 29/XII/1978, Mexico City.

Lindblom, Ch.E. (1965), *The Intelligence of Democracy: decision making through mutual adjustment,* Free, New York and Collier-Macmillan, London.

Liverman, D.M. (1993), 'Environment and Security in Mexico', in B.M. Bagley and S. Aguayo Quezada (eds), *Mexico: In Search of Security,* North-South Centre, University of Miami, Miami, pp. 211-246.

Local Government Act (LGA) (1985), *Local Government Act 1985,* Chapter 51, Her Majesty's Stationery Office (HMSO), London.

London Research Centre (LRC) (1993), *London Energy Study,* Prepared under the Commission of the European Communities' Urban and Regional Energy Management Programme, September, London Research Centre, London.

Lovell, L. and Johnson, D.L. (1994), 'The environmental crisis and direct social work practice', in M.D. Hoff and J.G. McNutt (eds), *The Global Environmental Crisis,* Avebury, Aldershot, pp. 199-218.

MacManus, P. (1996), 'Contested Terrains: Politics, Stories and Discourses of Sustainability', *Environmental Politics,* vol. 5, no. 1, Spring, Frank Cass, London, pp. 48-73.

Magnusson, W. (1986), 'Bourgeois Theories of Local Government', *Political Studies,* vol. 34, pp. 1-18.

Mannion, A.M. and Bowlby, S.R. (1992), 'Introduction', in A.M. Mannion and S.R. Bowlby (eds), *Environmental Issues in the 1990s,* Wiley, Chichester, pp. 3-20.

Marcondes, C. (1996), 'Latin America: Introduction', in P.L. McCarney (ed) (1996c), *The Changing Nature of Local Government in Developing Countries,* Centre for Urban and Community Studies, University of Toronto and International Office Federation of Canadian Municipalities, Toronto, pp. 223-226.

Marcou, G. (1993), 'New tendencies of local government development in Europe', in R.J. Bennet (ed), *Local Government in the New Europe,* Belhaven, London and New York, pp. 51-66.

Margulis, S. (1992), 'Back-of-the-Envelope Estimates of Environmental Damage Cost in Mexico', *Working Papers,* January, World Bank, Washington D.C.

Massey, D. and Meegan, R. (eds) (1985), *Politics and Method,* Methuen, London and New York.

Mawhood, P. (ed) (1993), *Local Government in The Third World,* Second Edition, Africa Institute of South Africa.

McCarney, P.L. (ed) (1996a), *Cities and Governance: New Directions in Latin America, Asia and Africa,* Centre for Urban and Community Studies, University of Toronto, Toronto.

McCarney, P.L. (1996b), 'Considerations on the Notion of 'Governance' – New Directions for Cities in the Developing World', in P.L. McCarney (ed), *Cities and Governance: New Directions in Latin America, Asia and Africa,* Centre for Urban and Community Studies, University of Toronto, Toronto, pp. 3-20.

McCarney, P.L. (ed) (1996c), *The Changing Nature of Local Government in Developing Countries,* Centre for Urban and Community Studies, University of Toronto and International Office Federation of Canadian Municipalities, Toronto.

McCarney, P.L. (1996d), 'Reviving Local Government: The Neglected Tier in Development', in P.L. McCarney (ed), *The Changing Nature of Local Government in Developing Countries,* Centre for Urban and Community Studies, University of Toronto and International Office Federation of Canadian Municipalities, Toronto, 5-30.

McCormick, J. (1991), *British Politics and the Environment,* Earthscan, London.

McCormick, J. (1995), *The Global Environmental Movement,* Second Edition, Wiley, Chichester.

Meadows, D.H. et al (eds) (1972), *The Limits to Growth,* Potomac Associates Books, New York.

Medina, S. and Quenel, P. (eds) (1993), *Technical report: Air Pollution and Health in large Metropolises,* June, Observatoire regional de sante d'Ile-de-France, Paris.

Miles, S.R. (ed) (1970), *Metropolitan Problems,* Methuen, Agincourt.

Mishler, E.G. (1986), *Research Interviewing,* Harvard University, Cambridge and London.

Morse, R.M. and Hardoy, J.E. (eds) (1992), *Rethinking the Latin American City,* Woodrow Wilson Center Press, Washington, D.C.

Murley, L. (ed) (1994), *Pollution Handbook 1994,* National Society for Clean Air and Environmental Protection (NSCA), Brighton.

National Environmental Technology Centre (NETCEN) (1995), *Air Pollution in the UK: 1993/1994,* AEA Technology, National Environmental Technology Centre, Culham.

National Society for Clean Air (NSCA) (1991), *Pollution Handbook, 1991,* National Society for Clean Air and Environmental Protection, Brighton.

Nava Escudero, C. (1992), *El Derecho de las Concesiones y Permisos en los Servicios y Obras Públicas en México. Estudio Constitucional y Administrativo,* unpublished thesis, First Degree in Law, Faculty of Law, Universidad Nacional Autónoma de México, Mexico City.

Nava Escudero, C. (1998), *Local Government Organisation in London and Mexico City: a comparative case-study of air quality management,* unpublished PhD thesis, London School of Economics and Political Science, London.

Navarro, B. (1993), *El Metro y sus Usuarios,* Universidad Autónoma Metropolitana (UAM), Mexico City.

Neira Alva, E. (1995), 'La Sustentabilidad de las Metrópolis Latinoamericanas', paper presented at *Foro del Ajusco II,* 31st May- 2nd June, Programa de las Naciones Unidas sobre el Medio Ambiente (PNUMA) y El Colegio de México, Mexico City.

Newman, P. and Thornley, A. (1997), 'Fragmentation and Centralisation in the Governance of London: Influencing the Urban Policy and Planning Agenda' *Urban Studies,* vol. 34, no. 7, pp. 967-988.

Niskanen, W.A. (1971), *Bureaucracy and representative government,* Aldine, Chicago.

Niskanen, W.A. (1973), *Bureaucracy, servant or master?,* Institute of Economic Affairs (IEA), London.

Norton, A. (1983), *The Government and Administration of Metropolitan Areas in Western Democracies,* Institute of Local Government Studies, University of Birmingham, Birmingham.

Nuccio, R.A. et al (1993), 'Mexico's Environment: Securing the Future', in B.M. Bagley and S. Aguayo Quezada (eds), *Mexico: In Search of Security,* North-south Centre, University of Miami, Miami, pp. 247-286.

Olson, M. (1965), *The Logic of Collective Action: Public Goods and Theory of Groups,* Harvard University, Cambridge.

Organisation for Economic Cooperation and Development (OECD) (1990), *Environmental Policies for Cities in the 1990s,* Organisation for Economic Cooperation and Development, Paris.

Organisation for Economic Cooperation and Development (OECD) (1995), *Control of Hazardous Air Pollutants in OECD Countries,* Organisation for Economic Cooperation and Development, Paris.

Organisation for Economic Cooperation and Development (OECD) (1996), *Innovative Policies for Sustainable Urban Development: The Ecological City,* Organisation for Economic Cooperation and Development, Paris.

O'Riordan, T. (1981), *Environmentalism,* Second Edition, Pion, London.

O'Riordan, I. and Weale, A. (1990), *Greening the Machinery of Government,* Discussion Paper No. 3, Friends of the Earth, London.

Osborne, D. and Gaebler, T. (1992), *Reinventing Government,* Plume, New York.

Ostro, B. et al (1995), 'Air Pollution and Mortality: Results from Santiago, Chile', *Working Papers,* May, World Bank, Washington D.C.

Ostrom, V. et al (1961), 'The Organisation of Government in Metropolitan Areas: a Theoretical Inquiry', *The American Political Science Review*, vol. 55, no. 4, December, pp. 831-842.

Ostrom, V. and Ostrom, E. (1971), 'Public Choice: A Different Approach to the Study of Public Administration', *Public Administration Review,* vol. 31, no. 2, March/April, American Society for Public Administration, pp. 203-216.

Pearson, Ch.S. and Pryor, A. (1978), *Environment: North and South,* Wiley, New York.

Pepper, D. (1984), *The roots of modern environmentalism,* Croom Helm, London.

Peterson, P. (1981), *City Limits,* University of Chicago, Chicago.

Pezzoli, K. (1991), 'Environmental conflicts in the urban milieu: the case of Mexico City' in D. Goodman and M. Redclift (eds), *Environment and development in Latin America,* Manchester University, Manchester, pp. 205-229.

Planning Act (1990), *Planning (Hazardous Substances) Act 1990,* Chapter 10, Her Majesty's Stationery Office (HMSO), London.

Quadri, G. (1993), 'La Contribución del Transporte a la Contaminación Atmosférica de la Ciudad de México', in Greenpeace Mexico, *El Transporte y la Contaminación,* Greenpeace Mexico, Mexico City, pp. 10-14.

Quadri, G. (1994), 'Economía, Sustentabilidad y Política Ambiental', in A. Yúnez-Naude (ed), *Medio Ambiente: problemas y soluciones,* El Colegio de México, Mexico City, pp. 21-61.

Quality of Urban Air Review Group (QUARG) (1993), *Urban Quality in the United Kingdom,* First Report of the Quality of Urban Air Review Group, January, Quality of Urban Air Review Group, London.

Redclift, M. (1984), *Development and the Environmental Crisis,* Methuen, London.

Redclift, M. and Goodman, D. (1991), 'Introduction', in D. Goodman and M. Redclift (eds), *Environment and development in Latin America: the politics of sustainability,* University of Manchester, Manchester, pp. 1-23.

Redclift, M. and Sage, C. (1994), 'Introduction', in M. Redclift and C. Sage (eds), *Strategies for Sustainable Development: local agendas for the Southern Hemisphere,* Wiley, Chichester, pp. 1-16.

Rees, A. (1991), *The Pocket Green Book: the Environmental Crisis in a Nutshell,* Zed Books, London and New Jersey.

Regan, D.E. (1972), 'London', in W.A. Robson and D.E. Regan (eds), *Great Cities of the World,* Third Edition, vol. II, Allen and Unwin, London, pp. 505-572.

Reglamento Interior de la Administración Pública del Distrito Federal (RIAPDF) (1995), *Reglamento Interior de la Administración Pública del Distrito Federal,* D.O. 15/IX/1995, Mexico City.

Restrepo, I. (ed) (1992), *La Contaminación Atmosférica en México: sus Causas y Efectos en la Salud,* Comisión Nacional de Derechos Humanos, Mexico City.

Rhodes, G. (1970), *The Government of London: the struggle for reform,* London School of Economics and Political Science, Weidenfeld and Nicolson, London.

Rhodes, G. (ed) (1972), *The new government of London: the first five years,* Greater London Group, London School of Economics, Weidenfeld and Nicolson, London.

Rhodes, R.A.W. (1980), 'Developed Countries', in D.C. Rowat (ed), *International Handbook on Local Government Reorganisation,* Aldwych, London, pp. 563-581.

Richardson, N.G. (1992), 'Canada', in R. Stren et al (eds), *Sustainable Cities: Urbanisation and the Environment in International Perspective,* Westview: Oxford, pp. 145-167.

Richardson, H.W. (1993), 'Efficiency and Welfare in LDC Meg-Cities', in J.D. Kasarda and A.M. Parnell (eds), *Third World Cities,* Sage, Newbury Park, pp. 32-57.

Ridley, T. and Travers, T. (1991), 'London Government and a Voice for Transport', in S. Glaister (ed), *Transport Options for London,* Greater London Group, London School of Economics and Political Science, London, pp. 161-176.

Robson, W.A. (1939), *The Government and Misgovernment of London,* Allen and Unwin, London.

Robson, W.A. and Regan, D.E. (eds) (1972), *Great Cities of the World,* Third Edition, vol. I and II, Allen and Unwin, London.

Rodríguez, A. and Winchester, L. (1996), 'The Challenges for Urban Governance in Latin America: Reinventing the Government of cities', in P.L. McCarney (ed), *Cities and Governance: New Directions in Latin America, Asia and Africa,* Centre for Urban and Community Studies, University of Toronto, Toronto, pp. 23-38.

Romieu, I. et al (1993), 'Air pollution and school absenteeism among children in Mexico City', in S. Medina and P. Quenel (eds), *Technical report: Air pollution and Health in large Metropolises,* June, Observatoire regional de sante d'Ile-de-France, Paris, pp. 298.

Romieu, I. et al (1995?), 'Effect of Urban Air Pollutants on Emergency Visists for Childhood Asthma in Mexico City', in M. Hernández Avila (ed), *Producción Bibliográfica en el Area de Salud Ambiental,* Centro de Investigaciones en Salud Pública, Cuernavaca.

Rose-Ackerman, S. (1983), 'Beyond Tiebout: Modeling the Political Economy of Local Government', in G.R. Zodrow (ed), *Local Provision of Public Services: The Tiebout Model After Twenty-Five Years,* Studies in Urban Economics, Academic, New York and London, pp. 55-83.

Rowat, D.C. (ed) (1973), *The government of federal capital,* University of Toronto: Toronto and Buffalo.

Rowat, D.C. (ed) (1980), *International Handbook, on Local Government Reorganisation,* Aldwych, London.

Royal Commission on Environmental Pollution (RCEP) (1971), *First Report,* Cmnd 4585, Her Majesty's Stationery Office (HMSO), London.

Royal Commission on Environmental Pollution (RCEP) (1974), *Fourth Report,* Cmnd 5780, Her Majesty's Stationery Office (HMSO), London.

Royal Commission on Environmental Pollution (RCEP) (1976), *Fifth Report,* Cmnd 6371, Chapter II, Her Majesty's Stationery Office (HMSO), London.

Royal Commission on Environmental Pollution (RCEP) (1984), *Tenth Report,* Cmnd 9149, Her Majesty's Stationery Office (HMSO), London.

Royal Commission on Environmental Pollution (RCEP) (1994), *Eighteenth Report,* Cm 2674, Her Majesty´s Stationery Office (HMSO), London.

Royal Commission on Local Government in Greater London (RCLG) (1960), *Royal Commission on Local Government in Greater London 1957-1960,* Herbert Report, Chapter II, Cmnd 1164, Her Majesty's Stationery Office (HMSO), London.

Royal Commission on Local Government in England (RCLG) (1969), *Redcliffe-Maud Report,* vol. 1, Cmnd 4040, Her Majesty´s Stationery Office (HMSO), London.

Ruck, S.K. and Rhodes, G. (1970), *The Government of Greater London,* Allen and Unwin, London.

Rydin, Y. (1997), 'Environmental Protection', in *Capital Choices: Essays on the new government for London,* a Report to the Association of London Government, October, Association of London Government (ALG), London.

Saint, A. (ed), (1989), *Politics and the People of London: The London County Council 1889-1965,* Hambledon, Ondon and Ronceverte.

Sandbrook, R. (1984), 'Opening the environmental debate', in D. Wilson (ed), *The Environmental Crisis. A Handbook for All Friends of the Earth,* Heinemann, London and Exeter, pp. 11-21.

Santos-Burgoa, C. and Rojas Bracho, L. (1992), 'Los Efectos de la Contaminación Atmosférica en la Salud', in I. Restrepo (ed), *La Contaminación Atmosférica en México: sus Causas y Efectos en la Salud,* Comisión Nacional de Derechos Humanos, Mexico City, pp. 205-250.

Satterthwaite, D. (1997), 'Sustainable Cities or Cities that Contribute to Sustainable Development?', *Urban Studies,* vol. 34, no. 10, pp. 1667-1691.

Sayer, A. (1992), *Method in Social Science: a realist approach,* Second Edition, Routledge, London and New York.

Sayer, A. and Morgan, K. (1985), 'A modern industry in a declining region: links between method, theory and policy', in D. Massey and R. Meegan (eds), *Politics and Method,* Methuen, London and New York, pp. 147-168.

Schteingart, M. (1989), 'The environmental problems associated with urban development in Mexico City', *Environment and Urbanization,* vol. 1, no. 1, April, pp. 40-50.

Schwartz, J. and Marcus, A. (1990), 'Mortality and Air Pollution in London: a Time Series Analysis', *American Journal of Epidemiology,* vol. 131, no. 1, The Johns Hopskins University School of Hygiene and Public Health, pp. 185-194.

Secretaría de Desarrollo Social (SEDESOL) (1993), *Informe de la Situación General en Materia de Equilibrio Ecológico y Protección al Ambiente 1991-1992,* Secretaría de Desarrollo Social and Instituto Nacional de Ecología, Mexico City.

Secretaría de Desarrollo Social (SEDESOL) (1994), *Informe de la Situación General en Materia de Equilibrio Ecológico y Protección al Ambiente 1993-1994,* Secretaría de Desarrollo Social and Instituto Nacional de Ecología, Mexico City.

Secretaría de Salud (Ssa) (1993), *Acciones desarrolladas para Evaluar el Efecto en la Salud, por la Contaminación Atmosférica en el Valle de México,* Sistema Nacional de Salud, Mexico City.

Secretariado Técnico Intergubernamental (STI) (1990), *Programa Integral Contra la Contaminación Amtosférica de la Zona Metropolitana de la Ciudad de México,* Secretariado Técnico Intergubernamental, Mexico City.

Self, P. (1962), 'The Herbert Report and the Values of Local Government', *Political Studies,* vol. X, no. 2, pp. 146-162.

Self, P. (1982), *Planning the Urban Region,* Allen and Unwin, London.

Self, P. (1993), *Government by the Market? The Politics of Public Choice,* Macmillan, Basingstoke and London.

Self, P. (1997), 'What's Happened to Administrative Theories?', *Public Policy and Administration,* vol. 12, no. 1, Spring, pp. 8-20.

Serageldin, I. et al (eds) (1995), *The Business of Sustainable Cities,* Second Annual World Bank Conference on Environmentally Sustainable Development, Environmentally Sustainable Development Proceedings Series No. 7, September 22-23, 1994, Washington D.C.

Serageldin, I. and Cohen, M.A. (eds) (1995), *The Human Face of the Urban Environment: a report to the development community,* Second Annual World Bank Conference on Environmentally Sustainable Development, Environmentally Sustainable Development Proceedings Series No. 5, September 10-21, 1994, Washington D.C.

Sharpe, L.J. (ed) (1993a), *The Rise of Meso Government in Europe,* Sage, London.

Sharpe, L.J. (1993b), 'The European Meso: An Appraisal' in L.J. Sharpe (ed), *The Rise of Meso Government in Europe,* Sage, London, pp. 1-39.

Sharpe, L.J. (ed) (1995a), *The Government of World Cities: the future of the metro model,* Wiley, Chichester.

Sharpe, L.J. (1995b), 'The Future of Metropolitan Government', in L.J. Sharpe (ed), *The Government of World Cities: the future of the metro model,* Wiley, Chichester, pp. 11-31.

Sharpe, L.J. (1995c), 'The Abolition of the Greater London Council: Is There a Case for Resurrection?', in L.J. Sharpe (ed), *The Government of World Cities: the future of the metro model,* Wiley, Chichester, pp. 111-130.

Sherlock, H. (1991), *Cities are good for us,* Paladin, London.

Smallwood, F. (1965), *Greater London,* Bobbs-Merrill, Indianapolis.

Soussan, J.G. (1992), 'Sustainable Development', in A.M. Mannion and S.R. Bowlby (eds), *Environmental Issues in the 1990s,* Wiley, Chichester, pp. 21-36.

South East Institute of Public Health (SEIPH) (1994), *Air Quality in London: the First Report of the London Air Quality Network,* February, South East Institute of Public Health, London.

Stewart, J. (1983), *Local Government: The conditions of Local Choice,* Institute of Local Government Studies, University of Birmingham, Allen and Unwin, London.

Stewart, J. (1986), *The New Management of Local Government,* Institute of Local Government Studies, University of Birmingham, Allen and Unwin, London.

Stewart, J. (1995), 'A Future for Local Authorities as Community Government', in J. Stewart and G. Stoker (eds), *Local Government in the 1990s,* Government Beyond the Centre, Macmillan, Basingstoke, pp. 249-267.

Stewart, J. and Stoker, G. (eds) (1995a), *Local Government in the 1990s,* Government Beyond the Centre, Macmillan, Basingstoke.

Stewart, J. and Stoker, G. (1995b), 'Fifteen Years of Local Government Restructuring 1979-94: An Evaluation', in J. Stewart and G. Stoker (eds), *Local Government in the 1990s,* Government Beyond the Centre, Macmillan, Basingstoke, pp. 191-209.

Stoker, G. (1991), *The Politics of Local Government,* Second Edition, Macmillan, Basingstoke.

Stoker, G. (1996a), 'Introduction: Normative Theories of Local Government and Democracy', in D. King and G. Stoker (eds), *Rethinking Local Democracy,* Government Beyond the Centre, Economic and Social Research Council, Macmillan, Basingstoke, pp. 1-27.

Stoker, G. (1996b), 'Governance as Theory: Five Propositions', Paper presented at *Enejeux Des Debates sur la Gouvernance,* Université de Lausanne, November 29-30, pp. 1-18.

Stoker, G. and Mossberger, K. (1995), 'The Post-Fordist Local State: The Dynamics of its Development' in J. Stewart and G. Stoker (eds), *Local Government in the 1990s,* Government Beyond the Centre, Macmillan, Basingstoke, pp. 210-227.

Stoker, G. and Young, S. (1993), *Cities in the 1990s,* Longman, Harlow.

Strauss, J. and Mainwaring, S.J. (1984), *Air pollution,* Edward Arnold, Baltimore.

Stren, R. (1992a), 'A Comparative Approach to Cities and the Environment', in R. Stren et al (eds), *Sustainable Cities: Urbanisation and the Environment in International Perspective,* Westview, Oxford, pp. 1-7.

Stren, R. (1992b), 'Conclusion', in R. Stren et al (eds), *Sustainable Cities: Urbanisation and the Environment in International Perspective,* Westview, Oxford, pp. 307-315.

Stren, R. (1996), 'Africa: Introduction', in P.L. McCarney (ed), *The Changing Nature of Local Government in Developing Countries,* Centre for Urban and Community Studies, University of Toronto and International Office Federation of Canadian Municipalities, Toronto, pp. 33-39.

Székely, F. (ed) (1978), *El Medio Ambiente en México y América Latina,* Nueva Imagen, Mexico City.

Tena Ramírez, F. (1970), *Derecho Constitucional Mexicano,* Tenth Edition, Porrúa, Mexico City.

Thomas, C. (1992), *The Environment in International Relations,* The Royal Institute of International Affairs, Earthscan, London.

Tickell, A. and Peck, J. (1996), 'The return of the Manchester men: men's words and men's deeds in the remaking of the local state', *Transactions, Institute of British Geographers,* New Series, vol. 21, no. 4, Dorchester, pp. 595-616.

Tiebout, Ch.M. (1956), 'A Pure Theory of Local Expenditures', *Journal of Political Economy,* vol. 64, pp. 416-424.

Travers, T. et al (1991a), *The Government of London,* London School of Economics and Political Science, Joseph Rowntree Foundation, London.

Travers, T. et al (1991b), *Planning,* Discussion Paper B, London School of Economics and Political Science, Greater London Group, London.

Travers, T. et al (1991c), 'Survey of Central Government Department and Agencies', in T. Travers et al, *The Government of London,* Discussion Paper G, August, London School of Economics and Political Science, Joseph Rowntree Foundation, London.

Travers, T. et al (1991d), *Transport,* Discussion Paper C, London School of Economics and Political Science, Greater London Group, London.

Travers, T. and Glaister, S. (1997), *Transport and the London Mayor,* The Greater London Group, London School of Economics and Political Science, London.

Travers, T. and Jones, J. (1997), *The New Government of London,* Joseph Rowntree Foundation, York.

Tullock, G. (1976), *The Vote Motive: an essay in the economics of politics,* Institute of Economic Affairs (IEA), London.

Turner, R.K. (1993), 'Sustainability: Principles and Practice', in R.K. Turner (ed), *Sustainable Environmental Economics and Management: Principles and Practice,* Belhaven, London, pp. 3-36.

United Nations (UN) (1984), *Statistics of Air Quality, Some Methods,* Conference of European Statisticians, Statistical Standards and Studies, no. 36, UN Statistical Commission and Economic Commission for Europe, Geneva.

United Nations (UN) (1993), *Earth Summit, Agenda 21: The United Nations Programme of Action from Rio,* UN Department of Public Information, New York.

United Nations Centre for Human Settlements - HABITAT (UNCHS) (1996a), *An Urbanizing World: Global Report on Human Settlements,* United Nations Centre for Human Settlements - HABITAT, Oxford University, Oxford.

United Nations Conference on Human Settlements - HABITAT (UNCHS) (1996b), 'Habitat II Spells out Road Map to the Future', *Habitat Debate,* September / December, vol. 2, no. 3 / 4, United Nations Centre for Human Settlements - HABITAT, Nairobi, pp. 1-5.

United Nations Environmental Programme (UNEP) (1993), *Environmental Data Report 1993-1994,* United Nations Environmental Programme (UNEP), World Resources Institute (WRI), Global Environmental Monitoring System (GEMS), Monitoring and Assessment Research Centre, Blackwell, Oxford.

Ward, P. (1990), *Mexico City,* Belhaven, London.

Ward, P. (1998), *Mexico City,* Second Edition, Wiley, Chichester.

Weir, F. (1993), *A Critique of UK Air Quality Strategy,* Friends of the Earth Paper to the IEHO Congress, September 13-16, The Institution of Environmental Health Officers, London.

Welch, R.V. (1997), 'Local Government and Sustainable Environment: European Perspectives', *European Environment,* vol. 7, no. 1, January-February, Wiley, Chichester, pp. 16-24.

White, R.R. (1993), *North, South and the Environmental Crisis,* University of Toronto, Toronto.

Wilson, D. (1984), 'The environmental crisis', in D. Wilson (ed), *The Environmental Crisis. A Handbook for All Friends of the Earth,* Heinemann, London and Exeter, pp. 1-10.

Wolman, H. (1995), 'Local Government Institutions and Democratic Governance', in D. Judge et al (eds), *Theories of Urban Politics,* Sage, London, pp. 135-159.

Wolman, H. (1996), 'Theories of Local Democracy in the United States', in D. King and G. Stoker (eds), *Rethinking Local Democracy,* Government Beyond the Centre, Economic and Social Research Council (ESRC) and Macmillan, Basingstoke, pp. 158-173.

World Bank (WB) (1994), *Making Development Sustainable,* Environment Department of the World Bank, Washington D.C.

World Commission on Environment and Development (WCED) (1987), *Our Common Future,* World Commission on Environment and Development, Oxford University, Oxford.

World Health Organisation (WHO) and United Nations Environment Programme (UNEP) (1992), *Urban Air Pollution in Megacities of the World,* Blackwell, Oxford.

World Resources Institute (WRI) (1992), *World Resources 1992-93,* World Resources Institute (WRI), United Nations Environment Programme (UNEP) and United Nations Development Programme (UNDP), Oxford University, New York and Oxford.

Young, K. and Garside, P.L. (1982), *Metropolitan London: politics and urban change 1837-1981,* Edward Arnold, London.

Zavala, S. (1992), *Orígenes del Distrito Federal,* Centro de Estudios de Historia de México, Condumsa, Mexico City.

Index